ANYONE CAN DO ARITHMETIC

BRIAN FLETCHER

Matador
5 Weir Road
Kibworth Beauchamp
Leicester LE8 0LQ, UK
Tel: 0116 279 2277
Email: books@troubador.co.uk
Web: www.troubador.co.uk/matador

ISBN 978 1848767 300

British Library Cataloguing in Publication Data.
A catalogue record for this book is available from the British Library.

Printed and bound in the UK by TJ International, Padstow, Cornwall

Matador is an imprint of Troubador Publishing Ltd

About the author

Brian Fletcher started his career as a Research Physicist specializing in accurate measurements of extremely low pressures. This was followed by many years teaching Mathematics up to University Entrance standard. The final five years of his career were spent teaching Information Technology to adults.

Now retired, he lives with his wife and dog in a small village in Wiltshire spending some of his time as a volunteer crew member and helmsman for a nearby canal charity and has recently become a Governor of the local primary school.

Acknowledgements

I am indebted to my son, Graham, for many helpful suggestions and his meticulous proof reading. My friend, Mo Woodgate, was the source of great encouragement when publication seemed a long way off. Thanks are also due to my wife, Sheila, for her patience when I was neglecting domestic duties while writing this book.

Brian Fletcher 2011

CONTENTS

INTRODUCTION

The motivation for writing this book was a report issued in February 2010 indicating that many primary school Mathematics teachers were unable to get correct answers to relatively simple problems. Apparently a few could not get the right answer to the calculation $2 \times 5 - 4$ while most could not correctly work out $2 \cdot 1\%$ of 400. This was from a relatively small sample of 155 but it is still rather alarming.

I do not believe that this is a purely modern phenomenon. In my experience and that of my peers it has long been the case that Mathematics was taught without any emphasis on understanding what you were doing. Following the rules and getting the right answer was the 'holy grail'.

In the late 1970s and early 1980s I was the Head of the Mathematics Department at a preparatory school where the Headmaster said to me one day 'there are only a certain number of questions and answers in Mathematics, so get the pupils to learn them all'.

I well remember my Arithmetic lessons in the late 1940s, the hours of chanting multiplication tables in order to get a star when I could recite one correctly. I was lucky enough to have a good memory and managed quite well but I do recall a problem with the answer to 6×9. If only a teacher had told me that multiplication works both ways and it was the same as 9×6 I would have made much faster progress. At the age of eleven we had games on a Wednesday afternoon. Before joining games we each had to do three multiplication sums (always containing hundreds, tens and units) correctly. I found the easiest way was to follow the rules even though I did not understand what I was doing. The slowest members of the class sometimes missed the whole games afternoon. This did not contribute to our enjoyment of Arithmetic.

Arithmetic is, in fact, a very easy subject, in spite of what many people may say. All that is necessary is to learn a few key answers (46 are sufficient) and remember the rules (there are not many of those). Then you can do Arithmetic but you will not understand it and you will find it very boring. When you have proper understanding you will find that you can do calculations in a way that suits you rather than slavishly copying the text book methods and further progress in Mathematics will be easier.

On the subject of calculation there has been criticism for many years of the fact that electronic calculators are widely used in schools by quite young children. Once the understanding of the principles has been grasped it seems to me rather silly not to take some of the hard work away. Aids to calculation have been used for thousands of years, the earliest being fingers. About 400 years ago a Scottish mathematician called John

Napier devised a set of rods called Napier's Bones to help with multiplication and division. The eastern world used the abacus. Just a few decades ago many shopkeepers would use a book of ready reckoner tables to work out prices. Mechanical and electrical adding machines were also in widespread use. Also used for multiplication and division were slide rules and tables of logarithms. The calculator is just the latest in a very long line of ways of making Arithmetic easier.

One of the difficulties in writing a book about Arithmetic is how much background knowledge to assume from the reader. I have made the decision (rightly or wrongly) to assume nothing more than the ability to read and to count. Some readers may find the beginnings of the early chapters rather trivial but please persevere. I would compare the development of a proper understanding of Arithmetic with building a house – if the foundations are not right then the whole building is at risk.

A question often posed is: 'Why learn Arithmetic when I have a calculator?' The simple answer is that, like a computer, a calculator has to receive the correct instructions in order to produce the correct answer. One of the decisions we all have to make these days is financial planning for retirement. The ability to do Arithmetic is vital to ensure the best choices are made. The wrong choice can lead to a much poorer life style in your retirement.

The subject of Mathematics covers a vast selection of topics and most of these form part of our everyday lives without being too obvious. Modern electronics would be impossible without Mathematics. Imagine, no I-pods, no television, no computers and many other things. Even Art, Music and Sculpture rely heavily on Mathematics.

Mathematics can be seen as part Science and part Art and it is reasonable to say it is the link between the two. Hence the importance of Arithmetic as it is the basic building block of Mathematics.

This book is almost completely about the understanding of numbers and the ways in which they may be combined to produce Arithmetic. It is invariably easier to do something with understanding than without it. There are many things in modern life that most of us do not understand, for example, if your DVD recorder breaks down we would call in an expert and pay for his services. Imagine the time and money wasted if we had to do this due to lack of ability to cope with Arithmetic.

The best way to read this book for the first time is in the order of the Chapters. If you skip around you may well be failing to understand something that relies on a previous Chapter. Please read on and allow me to prove that 'Anyone Can Do Arithmetic'.

CHAPTER ONE

A BRIEF HISTORY OF NUMBER

There was a young lad from the Humber
Having some trouble with number
Using paper and pen
He could count up to ten
But anything more made him slumber.

The ability to understand Arithmetic is likely to be helped somewhat by knowledge of the need for numbers and why we have the numerical systems of modern times.

The need for number was apparent many thousands of years ago, if only for comparison purposes. I will not bore the reader with the standard tales of notches carved in sticks and I will just give one up to date example involving something we all have an interest in: money! We all need to know how much money we have, and when we can see it in terms of notes and coins it is relatively easy. Some decades ago when most people were paid in cash, paid their bills with cash and stored any surplus under the mattress, it was easy to manage finances. Now, however, with salaries paid into banks and a lot of bill paying done by cheques, plastic and electronic transfer some representation using numbers (more correctly called numerals, but this difference only concerns mathematicians) and a system of Arithmetic is essential.

The first number system to be developed was the set of positive whole numbers, the ones we all use to count with. In various parts of the world different symbols and rules were used and many were discarded as their practical use was limited. One of the systems still used occasionally is the Roman Numeral system. This used I for one, V for five, X for ten, L for fifty, C for one hundred, D for five hundred and M for one thousand. The rules for intermediate numbers was to write these symbols one after the other and if a symbol was preceded by a larger symbol the two symbols were added but if a symbol was preceded by a smaller one then it was subtracted. *Rather complicated.* This meant that six was written as VI but four was written as IV. Larger numbers got very complicated. For example one thousand nine hundred and fifty six was written as MIMLVI. For anyone who wants a challenge try doing Arithmetic with Roman Numerals. (Don't panic – that was a joke.)

Our present system of numbers has survived only because of the addition of a symbol for zero and the idea of place value. We now use the symbols 0, 1, 2, 3, 4, 5, 6, 7, 8 and 9. Place value simply means that the same symbol can have different meanings depending on its position in the number.

With these ten symbols, when one of them is placed in the right hand position it represents units (or ones) and every time it is moved one place to the left its value gets ten times bigger. For example 2 represents units, but the 2 in 25 represents 2 tens and in 278 it means 2 hundreds and so on. This means the symbol for zero is essential in writing, say, two hundred as 200, the two zeros showing that the symbol 2 is in the hundreds place. We can now do the first piece of Arithmetic in this book. If we wish to multiply a whole number by ten we simply move all the digits one space to the left. This leaves the units space empty and this must be filled with a zero. This is the explanation for the well known rule 'to multiply a whole number by ten add a zero at the right hand end'. This method can be extended without limit, for example to multiply a whole number by one hundred place two zeros at the right hand end, to multiply by one thousand use three zeros and so on.

Using these ideas the number of symbols used is a matter of convenience. Deep in the heart of a computer only two symbols can be detected and we use 0 and 1. This means that numbers stored in a computer use more figures and Arithmetic takes more time but computers are so fast this is not important.

In later times it became apparent that there was a need for numbers smaller than zero. An easy example of numbers less than zero (negative numbers) is in the measurement of temperature. The weather forecasters use the Celsius scale of temperature which is based on $0°$ and $100°$ being the freezing and boiling points of water. We all know we have temperatures less than $0°$ (for example in the freezer part of a refrigerator) and these numbers have a minus sign to the left to indicate they are smaller than zero. For example, $-15°$ is a temperature $15°$ less than $0°$. These numbers are easier to understand when they are put on a number line (like a horizontal thermometer).

I have put a plus sign in front of the five right hand numbers to emphasise that they are positive. This sign is usually omitted unless essential. It is important to remember that when moving along this line from left to right you are going from small numbers to bigger ones and, of course, when moving from right to left you are going from large numbers to smaller ones. For example, it is obvious that 4 is a bigger number than 1 but it does seem strange at first to realise -4 is a smaller number than -1.

Our final set of numbers contains the ones that are missing on the number line. We only need to concern ourselves at the moment with the space between 0 and 1. These numbers are used when a unit has to be shared, or split into equal parts. They are called fractions and there are three common ways of representing these.

(a) Common or normal fractions. The technical name is vulgar fractions (this is not rude) but most people refer to them as just fractions.

These have two numbers separated either by a forward slash or a horizontal line. For example 1/4 or $\frac{1}{4}$

The second or lower of these numbers is more important. It is called the denominator but don't worry about the name. It tells us how many equal parts the unit has been split into. The first or upper number (the numerator, again don't worry about the name) tells us how many parts we require. The easiest way to understand this is with a diagram.

The diagram shows a unit split into three equal parts, two of which are shaded, so it is a pictorial representation of the fraction 2/3.

You will see later that one useful property of a fraction is that it can be represented in many ways. The fraction 3/4 is the same as 6/8, 9/12 etc. as can be seen from the following diagrams.

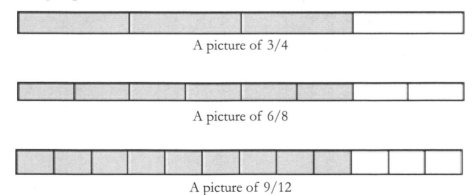

A picture of 3/4

A picture of 6/8

A picture of 9/12

One problem with fractions is that they can be difficult to compare, for example, deciding which of 3/7 and 4/9 is larger. This problem will be solved later. In Chapter 9 you will see how to change these numbers to decimal fractions.

(b) Decimal fractions. This system of fractions continues the idea of place value mentioned earlier. If moving a figure one place to the left increases its value by ten times then moving it one place to the right must reduce its value by ten times. So to the right of the tens place is the units, and the place to the right of that must be something ten times smaller than one and this is the tenths place. To use this and further places to the right we must separate the whole number part from the fraction part otherwise we would not know which place was which. This separator is the decimal point, strictly placed level with the middle of the line as shown here ·, but often a full stop is used.

So the fraction 0·7 is 7 tenths and the fraction 0·73 is 7 tenths and 3 hundredths.

(c) Percentage fractions. These fractions use the idea of a unit being split into 100 equal parts and the percentage is the number of parts required. A common use is to give examination results as a percentage, so 74 per cent, written as 74%, would be the equivalent of 74 marks out of 100. 74% could equally mean 148 marks out of 200 or 37 marks out of 50. (You will find out how this works in a later chapter.) Another well known use is to measure the rate of inflation; an annual inflation rate of 3% would mean that what would have cost £100 a year ago would now cost £103.

By combining whole numbers with fractions we have the basis of our present day number system.

35·2 means three tens, five units and two tenths,

$5^3/_4$ represents five units and three parts of a unit split into four.

Combining whole numbers with percentage fractions would mean we could have a number such as 206%. The use of these numbers will be explained in detail later.

CHAPTER TWO

ARITHMETIC WITH WHOLE NUMBERS

Whole numbers are fine, I can deal with those
They are quite easy as everyone knows.

Add, subtract and multiply
I am prepared to try,
But divide
I can't abide.

ADDING

The technical term 'operation' is used to mean a way of combining numbers, that is, adding, subtracting, multiplying and dividing. The easiest operation to do is adding. In the last resort, if you can count you can add. This is quite useful. As you read through this book you will see that all the other operations (subtracting, multiplying and dividing) can be done by adding. So, as long as you can count, you can do Arithmetic.

Addition is an operation that works both ways. In other words it does not matter which number comes first, so if you add 5 to 3 the answer is the same as adding 3 to 5. The sign for adding is the plus sign (+) so we can shorten the last sentence by saying 5 + 3 is the same as 3 + 5.

It is helpful to learn what are called number bonds for adding. This is the mathematician's way of saying you should know the result of adding any two single figure numbers. These are neatly summarized in the following table. Find the numbers you are adding in the first row and first column. The answer is where the row and column meet.

+	0	1	2	3	4	5	6	7	8	9
0	0	1	2	3	4	5	6	7	8	9
1	1	2	3	4	5	6	7	8	9	10
2	2	3	4	5	6	7	8	9	10	11
3	3	4	5	6	7	8	9	10	11	12
4	4	5	6	7	8	9	10	11	12	13
5	5	6	7	8	9	10	11	12	13	14
6	6	7	8	9	10	11	12	13	14	15
7	7	8	9	10	11	12	13	14	15	16
8	8	9	10	11	12	13	14	15	16	17
9	9	10	11	12	13	14	15	16	17	18

The numbers shown with a shaded background are not really necessary as these answers already appear in the rest of the table.

An addition sum is usually set out with the numbers to be added above each other with a ruled space for the answer as shown here:

$$2\ 3\ +$$
$$\underline{4\ 5}$$
$$\overline{}$$

The calculation proceeds by first adding the units column then adding the column to the left until all columns have been added so the complete sum would look like this:

$$2\ 3\ +$$
$$\underline{4\ 5}$$
$$\underline{6\ 8}$$

Now suppose the sum was:

$$2\ 7\ +$$
$$\underline{4\ 8}$$

The problem is that 7 + 8 (in the units column) is 15 and 15 is one ten and five units so the 1 must go in the tens column, but we will need the answer space when adding the tens. The solution is to write the 1 underneath the answer space of the tens column, remembering to include it when we add the tens. This is called carrying the 1 into the tens column. So we arrive at this stage:

$$
\begin{array}{r}
2\ 7\ + \\
\underline{4\ 8} \\
\underline{5} \\
1
\end{array}
$$

To complete the sum we now add 2 and 4, making 6 for the tens column and then adding the carried 1 giving 7. So the completed sum looks like this:

$$
\begin{array}{r}
2\ 7\ + \\
\underline{4\ 8} \\
\underline{7\ 5} \\
1
\end{array}
$$

Once you have understood what happens here it easy to extend the method to as many columns as you like. Three examples of complete sums involving more columns are shown here:

$$
\begin{array}{r}
2\ 6\ 8\ 3\ + \\
\underline{1\ 6\ 7} \\
\underline{2\ 8\ 5\ 0} \\
1\ 1
\end{array}
$$

$$
\begin{array}{r}
7\ 2\ + \\
\underline{6\ 9\ 5\ 5} \\
\underline{7\ 0\ 2\ 7} \\
1\ 1
\end{array}
$$

$$
\begin{array}{r}
6\ 8\ 4\ + \\
\underline{7\ 0\ 9} \\
\underline{1\ 3\ 9\ 3} \\
1\ 1
\end{array}
$$

Notice that if you needed any help in understanding this section, all the answers you wanted were in the table on page 6.

With practice adding more than two numbers will become as easy. With this sum:

$$7 +$$
$$4$$
$$\underline{2}$$
$$-$$

the problem is that the table only gives the answers for two numbers. There are two ways we can solve the problem. The long way is to do one sum for 7 + 4, getting the answer 11 and then doing another sum for 11 + 2, getting the final answer of 13. The short way is to use the result 7 + 4 =11 and then count on another 2 to arrive at 13. So the calculation would look like this:

$$7 +$$
$$4$$
$$\underline{2}$$
$$\underline{13}$$
$$1$$

A slight increase in difficulty would be this:

$$3\,7\,8 +$$
$$2\,7$$
$$\underline{1\,4\,9}$$

$$\overline{\qquad}$$

Your thinking should be 8 + 7=15 and a further 9 would make 24. This time you will be carrying a 2 into the tens column to arrive at this stage:

$$3\,7\,8 +$$
$$2\,7$$
$$1\,4\,9$$
$$\underline{\quad 4}$$
$$2$$

Now the tens column will be 7 + 2=9, 9 + 4=13 and the 2 to be carried makes 15.

$$3\,7\,8 +$$
$$2\,7$$
$$1\,4\,9$$
$$\underline{\quad 5\,4}$$
$$1\,2$$

Finally the hundreds column is completed; 3 + 1=4 and the 1 that is carried makes 5.

$$
\begin{array}{r}
3\ 7\ 8\ + \\
2\ 7 \\
\underline{1\ 4\ 9} \\
\underline{5\ 5\ 4} \\
1\ 2
\end{array}
$$

To end this section here are two additions for you to check to make sure you have understood everything so far.

$$
\begin{array}{r}
8\ 0\ 6\ + \\
4\ 7 \\
4\ 8\ 8 \\
\underline{3\ 2\ 9\ 4} \\
\underline{4\ 6\ 3\ 5} \\
1\ 2\ 2
\end{array}
$$

$$
\begin{array}{r}
6\ 5\ 8\ + \\
2\ 9 \\
8\ 7\ 5 \\
\underline{9\ 6\ 8} \\
\underline{2\ 5\ 3\ 0} \\
2\ 2\ 3
\end{array}
$$

If you need to build up your confidence you should try some addition sums of your own and check the answers with a calculator. (You may say 'Why not use a calculator all the time?' – Firstly, the aim of this book is to help you understand what you are doing and secondly, you may not always have a calculator with you).

One word of warning with calculators. Some cheap calculators do not always give the correct answer. A test for your calculator is to key in the following sum exactly as shown:

$$2 \times 3 - 3 \times 2 =.$$

If your calculator gives the answer 6, it will still work to check addition but do not trust it for other work until you have more knowledge of how to combine multiplying and subtracting. (The correct answer is 0). This will be explained further in Chapter 10.

SUBTRACTING

The main difference between adding and subtracting is that it does not work both ways. If you subtract 5 from 9 the answer is not the same as subtracting 9 from 5. The sign for subtracting is the minus sign (-) so 9 – 5 is not the same as 5 – 9. Notice how the numbers are reversed in the two ways of describing this calculation. You need to remember that 9 – 5 means the same as subtract 5 from 9.

The table of number bonds for subtracting is not very useful but it is shown below anyway.

−	0	1	2	3	4	5	6	7	8	9
0	0									
1	1	0								
2	2	1	0							
3	3	2	1	0						
4	4	3	2	1	0					
5	5	4	3	2	1	0				
6	6	5	4	3	2	1	0			
7	7	6	5	4	3	2	1	0		
8	8	7	6	5	4	3	2	1	0	
9	9	8	7	6	5	4	3	2	1	0

The empty spaces have not been filled in as we are only dealing with positive whole numbers in this chapter. If you check the answer to 9 – 5 you will see that you must find the first number in the left hand column, then follow the row until you are in the correct column for the second number.

A subtraction sum is set out in the same way as an addition sum. An easy one to start:

$$
\begin{array}{r}
2\,9 \; - \\
\underline{1\,3} \\
 \\
\end{array}
$$

Starting with the units column we can either use the table for 9 – 3, or, as promised, do it by working out what must be added to 3 to make 9. (You can do this by counting on your fingers if you wish. It is nothing to be ashamed of in spite of what you may have been told.) Either way the answer is 6.

$$
\begin{array}{r}
2\,9 \;- \\
\underline{1\,3} \\
\underline{6}
\end{array}
$$

Finally, the tens column, where 2 minus 1 is 1. So the complete sum is:

$$
\begin{array}{r}
2\,9 \;- \\
\underline{1\,3} \\
\underline{1\,6}
\end{array}
$$

The problems start when you try to take away a large single figure from a smaller one as in this sum:

$$
\begin{array}{r}
5\,2 \;- \\
\underline{1\,7} \\
 \\
\rule{1.5em}{0.4pt}
\end{array}
$$

We first wish to take 7 away from 2. Back to the number line, introduced on page 2.

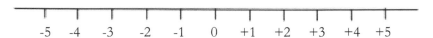

$$
\begin{array}{ccccccccccc}
-5 & -4 & -3 & -2 & -1 & 0 & +1 & +2 & +3 & +4 & +5
\end{array}
$$

If we start at +2 and move 7 places to the left we arrive at -5, which is correct. The sum could be completed in this way but there is a much easier method.

Remember 52 is 5 tens and 2 units and we can write this in a different way. Imagine someone was going to give you £52 (lucky you). I am sure you be would just as happy with four £10 notes and twelve £1 coins as you would be with five £10 notes and two £1 coins. So we write 52 as 4 in the tens column and 12 in the units. It is usually done like this:

$$
\begin{array}{r}
\overset{4}{\cancel{5}}\,{}^{1}2 \;- \\
\underline{1\,7} \\
 \\
\rule{1.5em}{0.4pt}
\end{array}
$$

This is what you may have heard called 'borrowing and paying back'.

So the units column is now 12 − 7 and by whatever method you use you should arrive at the answer 5. Carrying straight on with the tens column we work out 4 − 1 to be 3. The completed sum is now:

$$\begin{array}{r} 4 \\ \not{5}\,{}^1 2 \; - \\ \underline{1\;7} \\ \underline{3\;5} \end{array}$$

Until you become very confident it is a good idea to check your answer by addition. If you add your answer to what was taken away you should get back to the first number.

$$\begin{array}{r} 3\;5\;+ \\ \underline{1\;7} \\ \underline{5\;2} \quad \text{☑} \quad \text{☺} \\ 1 \end{array}$$

Notice how you 'carried' back the 1 that you 'borrowed',
   A slightly more difficult sum is this one:

$$\begin{array}{r} 3\;0\;2\; - \\ \underline{\;7\;5} \\ \underline{} \end{array}$$

In the units column 5 is bigger than 2 so we wish to borrow from the tens but there is only a zero there. We solve this problem by first borrowing 1 from the hundreds column:

$$\begin{array}{r} 2 \\ \not{3}\,{}^1 0\;2\; - \\ \underline{\;7\;5} \\ \underline{} \end{array}$$

We now have 2 in the hundreds and 10 in the tens column which we can now borrow from:

$$\begin{array}{r} 2 \quad 9 \\ \not{3}\,{}^1\not{0}\,{}^1 2\; - \\ \underline{\;7\;5} \\ \underline{} \end{array}$$

This looks a bit messy but we can now continue; 12 − 5 gives us 7 for the units, 9 − 7 is 2 for the tens and 2 − nothing is 2 for the hundreds.

$$\begin{array}{r} {\scriptstyle 2 \quad 9} \\ \cancel{3}\,{}^{\cancel{1}}\!\cancel{0}\,{}^{1}2 \ - \\ 7\ 5 \\ \hline 2\ 2\ 7 \end{array}$$

Checking by adding:

$$\begin{array}{r} 2\ 2\ 7 \ + \\ 7\ 5 \\ \hline 3\ 0\ 2 \qquad ☑ \quad ☺ \\ 1\ 1 \end{array}$$

This idea can be continued across as many columns as necessary.

At the start of this chapter it was stated that subtraction could be done by adding.

Before the introduction of electronic tills many shopkeepers would use adding to give correct change to a customer. Even though the following example concerns money (which is a mixture of whole numbers and decimal fractions and will be dealt with more fully in Chapter 7) it seems appropriate to explain the method here.

If you purchased an item for £2·38 and offered a £5 note in payment the shopkeeper would then have to work out £5·00 - £2·38 to find out how much change you required.

This is how it was done. He (or she) would give you a 2p piece while saying 'two pounds forty', then give you a 10p piece saying 'two pounds fifty', followed by a 50p piece and 'three pounds' and lastly two £1 coins while counting 'four, five' and you would have the correct change.

So the shopkeeper had found the correct answer to £5·00 - £2·38 by adding, in stages, the amount required to make £2·38 up to £5·00, and given change of 2p plus 10p plus 50p plus £2, which comes to £2·62.

This leads us to a clever way of subtracting from a string of zeros. If we need to calculate 1000 − 463 we would normally set the sum out like this:

$$\begin{array}{r} {\scriptstyle 9 \quad 9} \\ \cancel{1}\,{}^{\cancel{1}}\!\cancel{0}\,{}^{\cancel{1}}\!\cancel{0}\,{}^{1}0 \ - \\ 4\ 6\ 3 \\ \hline 5\ 3\ 7 \end{array}$$

If you look at the digits under the zeros you should be able to see that the right hand digit has to be made up to 10, while the others have to be made up to 9, in order to provide the correct answer.

We now have a very easy way of doing what may appear to be difficult subtraction sums.

10000 – 7354 can be calculated straightaway as 2646, (7 + 2 = 9, 3 + 6 =9, 5 + 4 = 9 and 4 + 6 ⁻ 10).

40000 – 7354 = 32646. Here you should realise that the leading 4 will have had a 1 'borrowed', leaving a 3 to start the answer.

Now, if you had offered a £10 note for your purchase costing £2·38, the answer to 10·00 - 2·38 can easily be worked out as 7·62.

The final stage in subtracting whole numbers is to take a large number away from a smaller one. Using the ideas shown so far this appears to be a very difficult thing to do but with a clever 'tweak' it can be made easy.

First we will try this with small numbers using the number line.

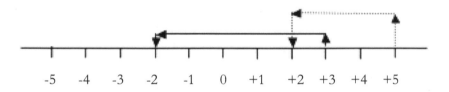

$$-5 \quad -4 \quad -3 \quad -2 \quad -1 \quad 0 \quad +1 \quad +2 \quad +3 \quad +4 \quad +5$$

The solid lines show the sum $3 - 5 = -2$ and the dotted lines show $5 - 3 = 2$.

So reversing the order gives the same number but without the minus sign. This is NOT a coincidence and it works with all numbers. (Try a few on your calculator if you are not convinced.)

If we wanted the answer to $27 - 98$, we would first work out $98 - 27$:

$$
\begin{array}{r}
9\,8 - \\
2\,7 \\
\hline
7\,1 \\
\end{array}
$$

and then supply the minus sign.

$$27 - 98 = -71$$

Once again it would be a good idea to try some subtraction sums and check your answers by adding and by calculator until you feel confident that you fully understand subtraction.

MULTIPLYING

Multiplying is actually a quick way of doing repeated addition, so it is another operation that works both ways. The sign for multiplication is a cross (×) and 5 × 4 gives the same answer as 4 × 5. The answer could be worked out just using addition in two ways:

```
    4 +      or      5 +
    4                5
    4                5
    4                5
    4               2 0
   2 0               2
    2
```

Multiplication is merely a way of shortening and tidying up this process.

You have probably had the task of learning something called 'times tables'. Often these had to be learned up to the '12 times table'. With the system of money we used before the 1970s 12 old pence made one shilling so the '12 times table' was considered a useful thing to know. With decimal money the '11 times table' and the '12 times table' are no longer needed and, of course, the '10 times table' is unnecessary as we found out in Chapter 1 how to multiply a number by 10.

The multiplication number bonds are shown in the table (as with addition, the lightly shaded numbers are repeated in the rest of the table).

×	0	1	2	3	4	5	6	7	8	9
0	0	0	0	0	0	0	0	0	0	0
1	0	1	2	3	4	5	6	7	8	9
2	0	2	4	6	8	10	12	14	16	18
3	0	3	6	9	12	15	18	21	24	27
4	0	4	8	12	16	20	24	28	32	36
5	0	5	10	15	20	25	30	35	40	45
6	0	6	12	18	24	30	36	42	48	54
7	0	7	14	21	28	35	42	49	56	63
8	0	8	16	24	32	40	48	56	64	72
9	0	9	18	27	36	45	54	63	72	81

This table works in the same way as the addition table. To multiply two numbers; find them in a row and a column; the answer is found where the row and column meet.

Of course if you multiply any number by zero the answer is zero, and if you multiply a number by 1 it stays the same. So the only answers that need to be learned are those that are heavily shaded. There are only 36 of them, much less than the 144 answers in all 12 'times tables'.

It is worth taking a little time to spot the many patterns in this table. For example all the 5s answers end in 5 or 0 and in the 9s answers the single figures add up to 9.

A simple multiplication sum (multiplying by a single figure) is set out in a similar way to an addition sum:

$$\begin{array}{r} 3\,7\times \\ \underline{4} \\ \underline{} \end{array}$$

At this stage it is worth taking a little time to check what we are doing.

We are trying to see what we get if we add four 37s together. Remember 37 is 3 tens plus 7 units so we want 4 lots of 7 units and 4 lots of 3 tens. We can get the answer with this information.

4 lots of 7 units are 28.

4 lots of 3 tens are 12 tens, which must be 120.

Add together 28 and 120 and the answer is 148.

One way of setting out this calculation would be:

$$\begin{array}{r} 4\times 7 = 2\,8\; + \\ 4\times 30 = \underline{1\,2\,0} \\ \underline{1\,4\,8} \end{array}$$

However there is a more compact layout:

$$\begin{array}{r} 3\,7\times \\ \underline{4} \\ 1\,4\,8 \\ 2 \end{array}$$

The thinking goes like this. 4 times 7 is 28. Write down 8 in the units space and carry 2 into the tens column. 4 times 3 is 12 (remember this really 4 times 30 = 120), add the 2 that was carried making 14 (this is really 120 + 20 = 140). Write 4 in the tens space and carry 1 into the hundreds column.

In this way all the figures are automatically placed in the correct column.

If you had any difficulty here try the calculation using money. Imagine you have three 10p coins and seven 1p coins and you wish to multiply this amount by 4.

Three 10p coins times 4 is twelve 10p coins which is 120p.

Seven 1p coins times 4 is 28p.

Add these two amounts together and you have 148p.

If you are still a little unsure, try some calculations of your own (only multiplying by a single figure at this stage) and check your answers before moving on.

When we multiply by a two figure number the setting out of the sum has to be expanded a little to make room for parts of the answer.

Consider what must be done to work out the answer to 43 × 26.

We must multiply 43 by 6:

$$
\begin{array}{r}
4\,3 \times \\
6 \\
\hline
2\,5\,8 \\
2\,1
\end{array}
$$

Then we multiply 43 by 20. We can get this answer if we multiply 43 by 2 and multiply the answer by 10:

$$
\begin{array}{r}
4\,3 \times \\
2 \\
\hline
8\,6
\end{array}
$$

and 86 × 10 is 860.

We must now add 258 and 860 to get the final answer:

$$
\begin{array}{r}
2\,5\,8 + \\
8\,6\,0 \\
\hline
1\,1\,1\,8 \\
1\,1
\end{array}
$$

If we now put these parts together we have the usual way of setting out a sum like this:

$$
\begin{array}{r}
4\,3 \times \\
2\,6 \\
\hline
2\,5\,8 \\
8\,6\,0 \\
\hline
1\,1\,1\,8
\end{array}
$$

At this stage you should get used to remembering the numbers to be carried in both the multiplication and the addition stages. So, this calculation would be built up in the following way:

$$
\begin{array}{r}
4\,3 \times \\
\underline{2\,6} \\
\end{array}
$$

6×3 is 18, so write 8 in the units space and remember to carry 1 into the tens.

$$
\begin{array}{r}
4\,3 \times \\
\underline{2\,6} \\
8 \\
\end{array}
$$

Now 6×4 is 24. Add the 1 being carried to get 25, then write 5 in the tens space and carry 2 into the hundreds (There are no more numbers for the hundreds so we can write the 2 in the correct space straight away).

$$
\begin{array}{r}
4\,3 \times \\
\underline{2\,6} \\
2\,5\,8 \\
\end{array}
$$

Start the next line with a zero as we are now multiplying by 20. 2×3 is 6 and this is written in the tens space.

$$
\begin{array}{r}
4\,3 \times \\
\underline{2\,6} \\
2\,5\,8 \\
6\,0 \\
\end{array}
$$

Now 2×4 is 8, which is placed in the hundreds.

$$
\begin{array}{r}
4\,3 \times \\
\underline{2\,6} \\
2\,5\,8 \\
\underline{8\,6\,0} \\
\end{array}
$$

Next the two parts of the answer must be added. $8 + 0$ is 8 so this goes in the units space in the final answer and $5 + 6$ is 11, so write 1 in the tens and carry 1 to the hundreds

```
    4 3  ×
      2 6
    2 5 8
    8 6 0
      1 8
```

Finally 2 + 8 is 10. Add the 1 to be carried giving 11. Write 1 in the hundreds space and carry a 1 to the thousands giving the final answer.

```
    4 3  ×
      2 6
    2 5 8
    8 6 0
  1 1 1 8
```

As multiplication works both ways we could have worked out 26 × 43 and expected to get the same answer.

```
    2 6  ×
      4 3
```

3 × 6 is 18, so write 8 in the units space and remember to carry 1 as before.

```
    2 6  ×
      4 3
        8
```

3 × 2 is 6. Plus 1 to be carried is 7, to be written in the tens space.

```
    2 6  ×
      4 3
      7 8
```

Start the next line with a zero as we are multiplying by 40. Then 4 × 6 is 24 so write 4 in the tens space and remember to carry 2.

```
    2 6  ×
      4 3
      7 8
      4 0
```

4×2 is 8. Adding 2 to be carried gives 10. So write 0 in the hundreds space and carry 1 into the thousands.

$$
\begin{array}{r}
2\ 6\ \times \\
\underline{4\ 3} \\
7\ 8 \\
\underline{1\ 0\ 4\ 0}
\end{array}
$$

Finally the adding, which should be straightforward by now.

$$
\begin{array}{r}
2\ 6\ \times \\
\underline{4\ 3} \\
7\ 8 \\
\underline{1\ 0\ 4\ 0} \\
\underline{1\ 1\ 1\ 8}
\end{array}
$$

Same answer!!! ☑ ☺

This process can also be shown on the diagram below:

	20	6
40	800	240
3	60	18

The four parts of the calculation, as before, are:

$40 \times 20 =$	800
$40 \times 6 =$	240
$3 \times 20 =$	60
$3 \times 6 =$	18
Adding the parts:	$\underline{1118}$

We have now seen the whole process for multiplying whole numbers. If the numbers have more than two digits we just continue following the same rules.

$$384 \times$$
$$59$$

9×4 is 36 so place a 6 in the units and carry 3 to the tens.

$$384 \times$$
$$59$$
$$6$$

9×8 is 72. $72 + 3$ is 75. Write 5 in the tens and carry 7 to the hundreds.

$$384 \times$$
$$59$$
$$56$$

9×3 is 27. $27 + 7$ is 34. Write 4 in the hundreds space and carry 3 to the thousands.

$$384 \times$$
$$59$$
$$3456$$

Start the next line with a zero, then 5×4 is 20, so write 0 in the tens and carry 2 to the hundreds.

$$384 \times$$
$$59$$
$$3456$$
$$00$$

5×8 is 40. $40 + 2$ is 42. Write 2 in the hundreds and carry 4 to the thousands.

$$384 \times$$
$$59$$
$$3456$$
$$200$$

5×3 is 15. $15 + 4$ is 19. Write 9 in the thousands and carry 1 to the ten thousands. Adding the two parts of the answer is now shown without further explanation.

$$
\begin{array}{r}
3\,8\,4 \times \\
5\,9 \\
\hline
3\,4\,5\,6 \\
\underline{1\,9\,2\,0\,0} \\
\underline{2\,2\,6\,5\,6}
\end{array}
$$

If you need to work out something like 268 × 347 a third line is needed in the working. By this stage I will assume that multiplying by 2 digits can be done and we will pick up the calculation from there.

$$
\begin{array}{r}
2\,6\,8 \times \\
3\,4\,7 \\
\hline
1\,8\,7\,6 \\
1\,0\,7\,2\,0
\end{array}
$$

Now we need the third line to record the answer to 268 × 300. We start the line with two zeros (the easy way of multiplying by 100) and then multiply by 3.

$$
\begin{array}{r}
2\,6\,8 \times \\
3\,4\,7 \\
\hline
1\,8\,7\,6 \\
1\,0\,7\,2\,0 \\
\underline{8\,0\,4\,0\,0}
\end{array}
$$

Finally add the three rows to get the answer.

$$
\begin{array}{r}
2\,6\,8 \times \\
3\,4\,7 \\
\hline
1\,8\,7\,6 \\
1\,0\,7\,2\,0 \\
\underline{8\,0\,4\,0\,0} \\
\underline{9\,2\,9\,9\,6}
\end{array}
$$

Now is the time for some practice, checking again with a calculator, until you feel confident enough to move on.

It may be useful to note that if you multiply the two units digits of the sum you will always get the units digit of the answer. You should be able to understand this by seeing where the zeros are placed in the working space. Also if you add together the number of digits of the two numbers in the sum then the number of digits in the answer will be this total or one less.

DIVIDING

Division is probably the least understood way of combining numbers for most people. It can be thought of in two ways which are best explained by examples.

36 divided by 9 can be understood to mean 'if you split 36 into 9 equal parts what is the value of each part?'.

It can also mean 'how many 9s do you need to make 36?'.

Those of you who have remembered the multiplication table will immediately know that 9×4 is 36 and so the answer is 4. So division can be thought of as the opposite of multiplication.

There are three signs for division and the sum above could be written as:

$$36 \div 9 \text{ or}$$

$$36/9 \text{ or}$$
$$\frac{36}{9}$$

The last two signs show a link between division and fractions to be explained later.

The answer table for division of whole numbers contains a lot of empty spaces. It is not a lot of help and is only included for completeness.

The empty spaces in the column under 0 are because we cannot divide a number by 0 using ordinary Arithmetic. If you consider the sum $2 \div 0$, for example, the question could be put as 'How many 0s do you need to make 2?'. The answer, of course, is that you could continue adding 0s together for as long as you liked and you would never arrive at 2. The other spaces cannot be filled in yet as the answers are fractions, and these will be explained later.

The row of zeros in the 0 row is because 0 cannot be divided into smaller parts.

÷	0	1	2	3	4	5	6	7	8	9
0		0	0	0	0	0	0	0	0	0
1		1								
2		2	1							
3		3		1						
4		4	2		1					
5		5				1				
6		6	3	2			1			
7		7						1		
8		8	4		2				1	
9		9		3						1

I have said that all Arithmetic can be done by adding and a simple division sum such as $8 \div 2$ is no exception. Remember the question is 'how many 2s will make 8?'. We can keep on adding 2s together until we reach an answer of 8 and then count the number of 2s.

$$
\begin{array}{ll}
2\ + & 1 \\
\underline{2} & 1 \\
4\ + & \\
\underline{2} & 1 \\
6\ + & \\
2 & \underline{1} \\
\underline{8} & \underline{4}
\end{array}
$$

So the answer is 4.

In the same way that multiplication is repeated addition, division can be done by repeated subtraction. So we could get the answer by starting with 8 and subtracting 2s as many times as we can.

A division sum is set out in a rather unusual way. This time the answer is placed ABOVE the horizontal line. A simple sum such as $8 \div 2$ looks like this:

$$2\,\overline{)\,8}$$

Notice the order of the numbers has been reversed and this is where we should think 'how many 2s are needed to make 8?' So the complete sum is:

$$\frac{4}{2)8}$$

In order to understand division the level of difficulty will be increased very slowly. Remember that division is repeated subtraction (seeing how many times we can take one number away from another).

Next we will try:

$$3)\overline{69}$$

If we were desperate we could keep subtracting 3 from 69 as many times as possible and count the number of 3s (or see how many 3s we needed to add together to make 69). Either process is fairly lengthy and can be made a lot simpler.

$$3)\overline{69}$$

69 is 6 tens and 9 units so if we can find out how many 3s make 60 and how many 3s make 9 we can simply add these two answers together. We first need to know how many 3s there are in 60. There are two 3s in 6 so there must be twenty 3s in 60. So 2 goes in the tens column of the answer space and 60 is taken away from 69 to see what is left.

$$
\begin{array}{r}
2 \\
3)\overline{6\,9} \\
\underline{6\,0} \\
9
\end{array}
$$

There are three 3s in 9 so 3 is written in the units column of the answer space and 9 is subtracted showing there is nothing left and we have completed the sum.

$$
\begin{array}{r}
2\,3 \\
3)\overline{6\,9} \\
\underline{6\,0} \\
9 \\
\underline{9} \\
\underline{0}
\end{array}
$$

In reality this sum would written in a much shorter form as shown:

25

$$\begin{array}{r} 2\,3 \\ \hline 3\,)\,6\,9 \end{array}$$

The reasoning would be this: 3 goes into 6 **exactly** 2 times so write 2 above the 6 and 3 goes into 9 **exactly** 3 times so write 3 above the 9. The words exactly are in bold because the reasoning has to be modified slightly if we do not have exact answers at each stage. We will see how this is done shortly.

We can check we have the right answer by seeing that 23 × 3 is in fact equal to 69.

The next stage is to see what we can do if we do not get an exact answer at the first part of the calculation.

The example is:

$$\begin{array}{r} \hline 3\,)\,7\,8 \end{array}$$

We need to know how many 3s there are in 70 this time. 3 goes into 70 more than 20 times but it is not a good idea to write anything in the units answer space yet so we will proceed as before putting 2 in the tens column of the answer space and subtracting 60.

$$\begin{array}{r} 2 \\ \hline 3\,)\,7\,8 \\ 6\,0 \\ \hline 1\,8 \end{array}$$

Now we have 18 left and 3 goes into 18 exactly 6 times and the calculation is complete.

$$\begin{array}{r} 2\,6 \\ \hline 3\,)\,7\,8 \\ 6\,0 \\ \hline 1\,8 \\ 1\,8 \\ \hline 0 \end{array}$$

Once again this sum is usually shortened to look like this:

$$\begin{array}{r} 2\ \ 6 \\ \hline 3\,)\,7\,{}^{1}8 \end{array}$$

The reasoning is as follows:

3 into 7 goes 2 times
Write 2 above the 7
3 times 2 is 6
7 minus 6 is 1 (this is in the tens column so it is worth 10 units)
Carry the 1 into the units making 18
3 goes into 18 exactly 6 times
Write 6 above the 8 giving the answer 26.

Check the answer. 3×26 is 78.

If the calculation was $79 \div 3$. the working would be as above except for the last three stages which would read:
Carry the 1 into the units making 19
3 goes into 19 6 times with 1 left over ($3 \times 6 = 18$)
Write 6 above the 9 and indicate what is left with the letter 'r' in front (r stands for remainder)

$$\frac{2\ 6}{3\,)\,7\,^1 9}\ \text{r}\ 1$$

To check the answer we must work out 3×36 and then add the remainder of 1.
    A slightly more difficult calculation would be $372 \div 4$.

$$4\,)\,\overline{3\ 7\ 2}$$

4 will not go into 3 so this must be carried into the next column. Remember the 3 represents 3 hundred which is the same as 30 tens. In this case we do not write anything above the 3. We do not start a whole number with 0 unless there is a good reason for doing so.

$$4\,)\,\overline{3\,^3 7\ 2}$$

From multiplication tables we know that 4×9 is 36, so 4×90 is 360 and this can be subtracted and 9 written in the tens column of the answer space.

$$\frac{9}{4\,)\,3\,^3 7\ 2}$$
$$\frac{3\ 6\ 0}{1\ 2}$$

27

Finally 4×3 is 12 so we write 3 in the units of the answer space and subtract the 12 to make sure there is nothing left.

$$
\begin{array}{r}
9\,{}^{1}3 \\
4\,)\,3\,{}^{3}7\,2 \\
\underline{3\ 6\ 0} \\
1\ 2 \\
\underline{1\ 2} \\
\underline{0}
\end{array}
$$

Again the answer can be checked by multiplying 93 by 4.

The shorter version is:

$$
\begin{array}{r}
9\ \ 3 \\
4\,)\,3\ {}^{3}7\ {}^{1}2
\end{array}
$$

And the thinking is:

4 will not go into 3. Carry the 3 to the next column. 4 goes into 37 nine times with 1 left over. Write 9 in the tens space. Carry 1 to the next column. 4 goes into 12 three times exactly. Write 3 in the units space.
  One more before we try dividing by larger numbers.

$$
6\,)\,\overline{1\ 2\ 3\ 6}
$$

6 will not go into 1 so this is carried into the next column.

$$
6\,)\,\overline{1\ {}^{1}2\ 3\ 6}
$$

6 will go into 12 exactly 2 times (remember, because of place value we are actually working out that 6 will go into 1200 exactly 200 times), so the next stage is:

$$
\begin{array}{r}
2\ \ \ \ \\
6\,)\,1\ {}^{1}2\ 3\ 6 \\
\underline{1\ 2\ 0\ 0} \\
3\ 6
\end{array}
$$

6 will not go into 3 so 0 is placed in the tens space to indicate that it is empty. 6 will go

into 36 exactly 6 times so the sum can be completed.

$$
\begin{array}{r}
2\,0\,6 \\
\overline{6)1\,^12\,3\,6} \\
1\,2\,0\,0 \\
3\,6 \\
\underline{3\,6} \\
0
\end{array}
$$

If we now multiply 206 by 6 we do get the result 1236 so our answer is correct.

Again the short calculation could be written like this:

$$
\begin{array}{r}
2\,0\ \,6 \\
\overline{6)1\,^12\,3\,^36}
\end{array}
$$

The associated reasoning is:

6 will not go into 1. 6 will go into 12 exactly 2 times so write a 2 in the hundreds column of the answer space.

6 will not go into 3 so we must place a 0 in the tens column to show it is empty. 6 will go into 36 exactly 6 times so place a 6 in the units to complete the sum. (The small carrying figures shown are optional. If you can remember them there is no need to write them.)

When dividing by numbers with two or more digits (sometimes called 'long division', which I am sure is enough to make anyone think it is more difficult) the only difference is that we cannot rely on multiplication tables for help. We have to rely on intelligent guesswork, which becomes easier with experience.

A fairly easy one to start with. 630 ÷ 18.

$$
\overline{18)630}
$$

18 clearly will not go into 6, but how many times will 18 go into 63?

This is where we have to 'guess'. If you can see that 18 × 2 is 36, it seems reasonable that we can try for a larger result, so try 18 × 3.

$$
\begin{array}{r}
1\,8\,\times \\
\underline{3} \\
5\,4
\end{array}
$$

If you try 18×4 you will get 72, which is too much, so the first stage can be done:

$$
\begin{array}{r}
3 \\
18\,)\,6\,3\,0 \\
\underline{5\,4\,0} \\
9\,0
\end{array}
$$

Remember one way of understanding this calculation is to realise that we are trying to take away as many 18s from 630 as possible. At this stage we have taken away 18 thirty times ($18 \times 30 = 540$) leaving us with 90. From above we know that $18 \times 4 = 72$ and it is not too difficult to see that 18×5 must be 90. ($72 + 18 = 90$)

The sum is now complete:

$$
\begin{array}{r}
3\,5 \\
18\,)\,6\,3\,0 \\
\underline{5\,4\,0} \\
9\,0 \\
\underline{9\,0} \\
0
\end{array}
$$

Check the answer:

$$
\begin{array}{r}
1\,8\,\times \\
\underline{3\,5} \\
9\,0 \\
\underline{5\,4\,0} \\
\underline{6\,3\,0}
\end{array}
$$

Correct answer!!! ☑ ☺

There is not much that can be done to shorten the working when dividing by larger numbers as the remainders arising at each stage are too big to be easily remembered and so we have to have the space to write them.

A slightly more difficult example is: $9176 \div 37$

$$
37\,)\,\overline{9\,1\,7\,6}
$$

37 will not go into 9 but will go into 91. $37 \times 2 = 74$ and $37 \times 3 = 111$, which is too

much. Now if $37 \times 2 = 74$ then $37 \times 200 = 7400$, so we subtract 7400 and record 2 in the hundreds column of the answer space:

$$
\begin{array}{r}
2 \\
37\overline{)9176} \\
\underline{7400} \\
1776
\end{array}
$$

37 will not go into 17 but will go into 177. $37 \times 4 = 148$ and $37 \times 5 = 185$, which is too much. Again if $37 \times 4 = 148$ then $37 \times 40 = 1480$, so we subtract 1480 and record 4 in the tens column of the answer space:

$$
\begin{array}{r}
24 \\
37\overline{)9176} \\
\underline{7400} \\
1776 \\
\underline{1480} \\
296
\end{array}
$$

The final stage is to find out how many 37s make 296. Knowing that $37 \times 10 = 370$, a shrewd guess would be to try 37×8 and sure enough it is 296 and the sum is now complete.

$$
\begin{array}{r}
248 \\
37\overline{)9176} \\
\underline{7400} \\
1776 \\
\underline{1480} \\
296 \\
\underline{296} \\
0
\end{array}
$$

If we check by multiplying we find that 248×37 is 9176.

This calculation is slightly shortened by omitting the numbers that are crossed out. The zeros are only there to remind us that we are really subtracting tens and hundreds and the six is just part of the original sum.

The layout and reasoning would normally be as follows:

```
  _____
37)9176
```

How many 37s make 91?

$37 \times 2 = 74$ so subtract 74 from 91 and place 2 in the answer space above the 1.

```
      2
  _____
37)9176
  74
  17
```

'Bring down the next figure'. This is what happens when you subtract 0:

```
      2
  _____
37)9176
  74
  177
```

How many 37s make 177?

$37 \times 4 = 148$ so subtract 148 from 177 and place 4 in the answer space above the 7.

```
     24
  _____
37)9176
  74
  177
  148
  29
```

'Bring down the next figure':

```
     24
  _____
37)9176
  74
  177
  148
  296
```

How many 37s make 296?

$37 \times 8 = 296$ so subtract 296 and place 8 in the answer space above the 6.

```
            2 4 8
3 7 ) 9 1 7 6
        7 4
        1 7 7
        1 4 8
          2 9 6
          2 9 6
            0
```

The calculation is now complete and probably looks the way older readers were taught at school (without too much explanation).

If you find this difficult there is an easier, but slightly longer way of solving the problem involving subtracting easy multiples of 37.

	Number of 37s subtracted
9 1 7 6	
3 7 0 0	1 0 0
5 4 7 6	
3 7 0 0	1 0 0
1 7 7 6	
3 7 0	1 0
1 4 0 6	
3 7 0	1 0
1 0 3 6	
3 7 0	1 0
6 6 6	
3 7 0	1 0
2 9 6	
3 7	1
2 5 9	
3 7	1
2 2 2	
3 7	1
1 8 5	
3 7	1
1 4 8	
3 7	1
1 1 1	
3 7	1
7 4	
3 7	1
3 7	
3 7	1
0	Total = 2 4 8

Once again practice as much as you need to feel confident before moving on.

CHAPTER THREE

ARITHMETIC WITH VULGAR FRACTIONS

Fractions are bits, sometime quite small
Why they are vulgar I know not at all.

ADDING

Before embarking on this topic we need to know a little more about vulgar fractions (referred to in the rest of this chapter as just fractions).

In Chapter 1 it was pointed out that the two numbers making up a fraction have different meanings. The lower number tells us how many parts a unit has been divided into, while the upper number tells us how many of these parts are needed to make the fraction. So the lower number refers to the type of fraction and the upper one is simply a number.

Adding two fractions where the lower numbers are the same is fairly easily understood when the sum is posed in words. For example, what is one fifth plus three fifths (1/5 + 3/5). The total number of fifths is 1 + 3 = 4, so the answer is 4/5.

This poses a problem when adding fractions with different lower numbers as we are trying to add two different types of fraction. A similar problem occurs when we try to add different types of money, for example what is £23 + $62? To do this calculation we need to know the exchange rate between the British pound and the American dollar. Once we then have the sum entirely in pounds or dollars it is easy.

Fortunately with fractions the conversion is much easier.

On page 3 you saw the idea of a fraction being written in different ways:

So   3 is the same as 6 and this is the same as 9
        4   8   12

Notice that to progress from 3/4 to 6/8 both the top and bottom numbers are multiplied by 2, and to move on to 9/12 we multiply both numbers by 3.

This can be continued as much as you like so 12/16 and 15/20 are all the same as 3/4. In fact, as long as both numbers in a fraction are multiplied by the same number the fraction retains its value. The reason this works is that 3/3, for example, means 3 ÷ 3 which is equal to 1, and multiplying a number by 1 does not change anything. This also means that if it is possible to divide both parts of a fraction by the same number then the fraction can be written in a simpler form: i.e. the fraction 15/25 means exactly the same as 3/5. (Both 15 and 25 can be divided by 5)

Now to add two fractions where the lower numbers are different we must use the ideas of the last paragraph to make them the same.

An easy one to start with:

$$\frac{1}{2} + \frac{1}{4}$$

It is easy to see here that 1/2 can be replaced by 2/4 and the sum is now:

$$\frac{2}{4} + \frac{1}{4} = \frac{3}{4}$$

A slightly more difficult sum would be:

$$\frac{1}{2} + \frac{1}{3}$$

We now need to make the two lower numbers the same by multiplying. If we multiply 1/2 by 3/3 we get 3/6 and multiplying 1/3 by 2/2 gives 2/6. Now the sum becomes:

$$\frac{3}{6} + \frac{2}{6} = \frac{5}{6}$$

This technique will **always** work, so we can multiply both numbers of the first fraction by the bottom number of the second and multiply both numbers of the second fraction by the bottom number of the first. This explanation is a bit long-winded but a few examples should make it clear.

If we try:

$$\frac{2}{5} + \frac{1}{3}$$

the first fraction is multiplied by 3/3 and the second fraction is multiplied by 5/5 and the sum becomes:

$$\frac{6}{15} \quad + \quad \frac{5}{15} \quad = \quad \frac{11}{15}$$

One more:

$$\frac{3}{4} \quad + \quad \frac{1}{5}$$

$$\frac{15}{20} \quad + \quad \frac{4}{20} \quad = \quad \frac{19}{20}$$

The 'long-winded' explanation on the previous page will always work, but we sometimes get an answer which is perfectly correct but is not in its simplest form.

An example of this is:

$$\frac{1}{4} \quad + \quad \frac{1}{6}$$

Using our previous technique this sum becomes:

$$\frac{6}{24} \quad + \quad \frac{4}{24} \quad = \quad \frac{10}{24}$$

We already know that if the two numbers of a fraction can be divided by the same number we can do this without changing the value. In this case both numbers can be divided by 2 so it would be better if we gave the answer as 5/12.

Remember that when adding fractions we are trying to make the lower number the same in each case by multiplying. When we multiply 4 by different numbers we get what are called multiples of 4. The first few multiples of 4 are shown below. (The shading is explained immediately after the diagrams).

4	8	12	16	20	24	28	32	36	40	44	48

Now the first few multiples of 6:

6	12	18	24	30	36	42	48	54	60	66	72

The numbers with shaded backgrounds appear in both sets of numbers and are multiples of 4 and multiples of 6. These are called common multiples of 4 and 6. Using the smallest of these common multiples when adding fractions keeps all the numbers smaller so it is the smart thing to do if you can spot it. So, in a perfect world the previous sum would be:

$$\frac{3}{12} + \frac{2}{12} = \frac{5}{12}$$

Don't worry if you cannot do this yet. It becomes easier with practice. You can always simplify your answer at the end.

See if you can follow this one:

$$\frac{3}{8} + \frac{1}{12}$$

Multiply the first fraction by 3/3 and the second by 2/2.

$$\frac{9}{24} + \frac{2}{24} = \frac{11}{24}$$

We can save some writing when adding fractions. We know we are going to make the lower numbers the same and add the top numbers so there is no need to repeat the lower numbers. The last sum could have been written as:

$$\frac{9 + 2}{24} = \frac{11}{24}$$

Just one more sum now to show what we do if the answer is bigger than 1:

$$\frac{5}{8} + \frac{7}{12}$$

We know that the denominator needs to be 24 so we write this as:

$$\frac{15 + 14}{24} = \frac{29}{24}$$

The answer in this form is called an improper fraction. A fraction is smaller than 1 and so the top number would be smaller than the bottom number. Now 24/24 is the same as 1 and we then have 5/24 left so the usual way of writing this answer is:

$$1\frac{5}{24}$$

This is called a mixed number, (more of this later).

If you wish to practice adding fractions and check your answers you will need a calculator that deals with fractions. The fraction button on a calculator usually looks something like a^b/c. The last calculation would be done by pressing the following buttons:

$$5 \; a^b/c \; 8 + 7 \; a^b/c \; 12 =$$

And the answer will be shown as: 1   5   24.

Notice that the calculator gives the answer in the simplest form. If you had obtained an answer of 39/52 to some calculation and wondered if it could be made simpler then enter 39 a^b/c 52 = and your calculator will show 3   4 (3/4).

SUBTRACTING

You will be pleased to know that we have done all the hard work for this section when learning how to add fractions. The lower numbers must be made the same for the same reasons that were explained when adding and then the top numbers must be subtracted instead of added.

So:

$$\frac{1}{2} \quad - \quad \frac{1}{4}$$

becomes:

$$\frac{2 \quad - \quad 1}{4} \quad = \quad \frac{1}{4}$$

and

$$\frac{4}{5} \quad - \quad \frac{2}{3}$$

becomes:

$$\frac{12 \quad - \quad 10}{15} \quad = \quad \frac{2}{15}$$

and, lastly:

$$\frac{7}{8} \quad - \quad \frac{5}{12}$$

becomes:

$$\frac{21 \quad - \quad 10}{24} \quad = \quad \frac{11}{24}$$

I hope you spotted that 24 is smallest multiple of 8 and 12.

Remember all subtraction sums can be checked by adding.

$$\frac{11 \quad + \quad 10}{24} \quad = \quad \frac{21}{24} \quad = \quad \frac{7}{8}$$

MULTIPLYING

In order to understand the way in which fractions are multiplied it helps to examine the meaning of the word 'of'. If you were buying loaves OF bread, priced at £1.35 each, and at the baker's shop you pointed at your selection and said 'I want 3 OF these' the baker would have to work out 3 × £1.35 for your bill. (Of course, in reality, the electronic till would do the calculation.) So there is an obvious link between the word 'of' and multiplication. Similarly, if you wanted to buy 1/2 of a metre of wood priced at £1.24 a metre, the calculation is 1/2 times £1.24 which is £0.62.

As £1.24 divided by 2 is also £0.62 you may be wondering if multiplying by 1/2 is the same as dividing by 2. The answer is a definite yes.

The more curious may now be pondering whether dividing by 1/2 is the same as multiplying by 2. Again, the answer is yes, but more of this in the next section.

As always we will start with a very simple sum and then build up to more difficult ones.

$$\frac{1}{2} \quad \times \quad \frac{1}{2}$$

Remember we want to work out 1/2 of 1/2. A fraction diagram may help.
This is our diagram for 1/2. To get the answer we must split the shaded box into two equal parts but to find out what type of fraction we have, the complete rectangle must consist of equal size boxes.

We now have 4 boxes, so each box represents 1/4. Our original 1/2 is now 2/4 and we are looking for 1/2 of 2/4 which must be 1/4. So we have our answer.

$$\frac{1}{2} \quad \times \quad \frac{1}{2} \quad - \quad \frac{1}{4}$$

It seems as if the 4 in the answer comes from multiplying 2 and 2. The question now is to see if this always works.

We will now try:

$$\frac{1}{2} \quad \times \quad \frac{1}{3}$$

Now to find 1/2 of 1/3 we must divide each of the boxes into 2 equal parts.

We now find 1/3 is shown as 2/6 and we want half of this, which is 1/6.

$$\frac{1}{2} \quad \times \quad \frac{1}{3} \quad = \quad \frac{1}{6}$$

Again we have multiplied 2 and 3 to get the 6 of the answer.

One more to help convince you:

$$\frac{1}{2} \quad \times \quad \frac{1}{7}$$

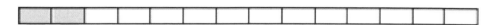

Divide each box into 2 equal parts:

We now have 1/7 shown as 2/14 and hence the answer:

$$\frac{1}{2} \quad \times \quad \frac{1}{7} \quad = \quad \frac{1}{14}$$

The final example:

$$\frac{1}{3} \quad \times \quad \frac{1}{5}$$

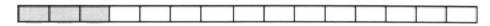

Divide each box into 3 equal parts:

1/5 is now shown as 3/15 and we want 1/3 of this which is 1/15.

The first part of the rule for multiplying fractions is now established. Provided the top numbers are both 1 we simply multiply the bottom numbers.

To get the rest of the rule we will look at variations of the last sum.

First:

$$\frac{1}{3} \quad \times \quad \frac{2}{5}$$

2/5 is twice as big as 1/5 so the answer here must be twice as big as the previous answer, so:

$$\frac{1}{3} \quad \times \quad \frac{2}{5} \quad = \quad \frac{2}{15}$$

Next:

$$\frac{1}{3} \quad \times \quad \frac{3}{5}$$

3/5 is three times as big as 1/5 so the answer is three times as big:

$$\frac{1}{3} \quad \times \quad \frac{3}{5} \quad = \quad \frac{3}{15}$$

But both 3 and 15 can be divided by 3 to give a simpler answer of 1/5.
    If we try

$$\frac{2}{3} \quad \times \quad \frac{3}{5}$$

You should see that 2/3 is twice as much as 1/3 so the answer will be twice as big.

$$\frac{2}{3} \quad \times \quad \frac{3}{5} \quad = \quad \frac{6}{15}$$

This answer can also be simplified by dividing 6 and 15 by 3 to give 2/5.

These examples show that we should multiply the top numbers together as well as the bottom numbers.

In the last two sums we have been able to simplify the answer by dividing the top and bottom numbers by the same number. It is possible to do this before working out the answer. If any number on top and any number on the bottom can be divided by the same number then this can be done before finding the answer. In the last example the 3 on top and the 3 underneath can both be divided by 3 leaving a 1, as shown below:

$$\frac{2}{^{1}3} \quad \times \quad \frac{^{1}3}{5} \quad = \quad \frac{2}{5}$$

This process is known as cancelling. We have now worked out the rule for multiplying fractions. It is:

first do any possible cancelling, then multiply the top numbers and multiply the bottom numbers.

Here are some examples to show how it is done.

$$\frac{2}{3} \quad \times \quad \frac{6}{7}$$

The 6 on top and the 3 underneath can both be divided by 3, so the cancelling is done:

$$\frac{2}{^13} \quad \times \quad \frac{^26}{7}$$

The answer now comes by multiplying 2 by 2 giving 4 as the numerator, and multiplying 1 by 7 making 7 for the denominator.

$$\frac{2}{^13} \quad \times \quad \frac{^26}{7} \quad = \quad \frac{4}{7}$$

One more:

$$\frac{6}{7} \quad \times \quad \frac{5}{12}$$

The 6 on top and the 12 underneath can both be divided by 6, arriving at the answer as shown:

$$\frac{^16}{7} \quad \times \quad \frac{5}{^212} \quad = \quad \frac{5}{14}$$

Sometimes the cancelling can be done twice as in the last example for this section.

$$\frac{5}{6} \quad \times \quad \frac{12}{25}$$

The 5 on top and the 25 below can both be divided by 5, and the 12 on and the 6 below can both be divided by 6. When this is done the calculation and the answer look like this;

$$\frac{^1\cancel{5}}{^1\cancel{6}} \quad \times \quad \frac{^2\cancel{12}}{^5\cancel{25}} \quad = \quad \frac{2}{5}$$

If you don't spot the cancelling every time it just means your answer has bigger numbers which can then be simplified.

DIVIDING

The rule that we will end up with for dividing fractions seems rather strange if it is just presented without explanation so we will build up to it in stages.

First we will just consider dividing whole numbers by fractions like 1/2, 1/3, etc.

$$\frac{1}{1} \quad \div \quad \frac{1}{2}$$

As this chapter is about fractions the number 1 is shown as the fraction 1/1. This, of course, is the same thing as 1 and shows how we can write a number in different ways if it is convenient.

The question now is how many 1/2s do we need to make 1 and the answer, of course, is 2 as 1/2 is simply 1 divided into 2 equal parts. So the first division of fractions is done:

$$\frac{1}{1} \quad \div \quad \frac{1}{2} \quad = \quad \frac{2}{1}$$

Once again, the answer 2 is written as the fraction 2/1.

Notice we could have achieved the same answer if we multiplied by 2/1 as shown:

$$\frac{1}{1} \quad \times \quad \frac{2}{1} \quad = \quad \frac{2}{1}$$

This poses the question: Is dividing by 1/2 the same as multiplying by 2? It does seem likely as if you require 1/2 OF something (multiplying) you divide it by 2. So it seems reasonable to suppose that if you require 2 of something you could divide by 1/2. The only possible reason to do this peculiar sum is to build up our method for dividing fractions.

Try this one:

$$\frac{1}{1} \div \frac{1}{3}$$

With the same reasoning as before, we ask how many 1/3s do we need to make 1? The answer, of course, is 3.

Again, switching to multiplying works:

$$\frac{1}{1} \times \frac{3}{1} = \frac{3}{1}$$

It should be clear this will work with 1/4, 1/5, 1/6 and so on.

Now we will try a slightly more difficult example:

$$\frac{1}{1} \div \frac{2}{3}$$

2/3 is twice as big as 1/3, so when we ask how many 2/3s do we need to make 1 we should get half of the previous answer, i.e. 3/2.

Our method (changing to multiplying and turning the second fraction upside down) still works.

$$\frac{1}{1} \times \frac{3}{2} = \frac{3}{2}$$

Increasing the difficulty again:

$$\frac{5}{1} \div \frac{1}{3}$$

The question to ask is how many 1/3s do we need to make 1? Three 1/3s make 1 so we need 5 times as many to make 5, so the answer is 15, as shown:

$$\frac{5}{1} \div \frac{1}{3} = \frac{15}{1}$$

Again the same answer is obtained by turning the second fraction upside down and multiplying:

$$\frac{5}{1} \quad \times \quad \frac{3}{1} \quad - \quad \frac{15}{1}$$

A little harder:

$$\frac{5}{4} \quad \div \quad \frac{1}{3}$$

The only difference from the last sum is that the first fraction has been divided by 4, so the answer must be divided by 4, and:

$$\frac{5}{4} \quad \div \quad \frac{1}{3} \quad = \quad \frac{15}{4} \quad = \quad 3\frac{3}{4}$$

This could equally have been achieved by multiplying as before:

$$\frac{5}{4} \quad \times \quad \frac{3}{1} \quad = \quad \frac{15}{4} \quad = \quad 3\frac{3}{4}$$

The pressure is now on for the final example, which I will explain in detail.

$$\frac{3}{7} \quad \div \quad \frac{4}{5}$$

Remember, we are asking how many 4/5s do we need to make 3/7.

Stage 1: How many 1/5s do we need to make 1?

Answer   5

Stage 2: How many 1/5s do we need to make 3?

Answer 5×3 =   15

Stage 3: How many 4/5s do we need to make 3? 4/5 is four times the size of 1/5 so we need 1/4 as many.

Answer $\qquad \dfrac{5 \times 3}{4} = \dfrac{15}{4}$

Stage 4: How many 4/5s do we need to make 3/7? We now have 1/7 of the original three so we must divide the answer by 7.

Answer $\qquad \dfrac{5 \times 3}{4 \times 7} = \dfrac{15}{28}$

The completed sum is now:

$$\dfrac{3}{7} \div \dfrac{4}{5} = \dfrac{15}{28}$$

This confirms that turning the second fraction upside and multiplying will work.

$$\dfrac{3}{7} \times \dfrac{5}{4} = \dfrac{15}{28}$$

The understanding of how to divide fractions causes so many problems I feel it is worth offering another explanation.

Consider the last sum again:

$$\dfrac{3}{7} \div \dfrac{4}{5}$$

We can multiply the first fraction by 5/5 and the second one by 7/7 without changing their values, so the sum is:

$$\dfrac{15}{35} \div \dfrac{28}{35}$$

Both fractions are now of the same type so we just need to know how many 28s we need to make 15. This is just dividing 15 by 28 which we can write as 15/28.

$$\dfrac{3}{7} \div \dfrac{4}{5} = \dfrac{15}{28}$$

Same answer (thank goodness!). As long as you understand one of the explanations that is sufficient.

CHAPTER FOUR

ARITHMETIC WITH DECIMAL FRACTIONS

Ode to a decimal point

Whole numbers to the left of me, fractions to the right
I sit in the middle, wedged in tight.

First some comments about how decimal fractions (from here on just known as decimals) are written. In everyday life it is not normal to start a whole number with a zero. For example, it would be unusual to write 015 when we meant 15. However there are exceptions.

One is the numbering of bank accounts. Since 2002 bank account numbers have had to be '8 numeric characters with leading zeros added if necessary'. So, if you have had a bank account for many years and its original number was 582296 it will now be 00582296. (If this happens to be a real bank account number I must apologise to the owner even though I could not know who it is.)

The other common exception is when writing decimals. 0·93 is the correct way to write ·93. The reason for this is probably historic. In the days when many documents were hand written, the zero in the units column was to emphasise that a decimal point was coming next. Without this someone quickly scanning the document may have misread ·46 as 46.

ADDING and SUBTRACTING

These two operations can be considered together as the only difference between whole numbers and decimals is the way in which the numbers are lined up. With whole numbers we had to make sure the units column of each number was in line, whereas with decimals we keep the decimal points in line.

Just a few examples of each operation should suffice.

$$0·356 + 0·213$$

This is set out in the usual way and added column by column from the right:

$$0 \cdot 3\,5\,6\ +$$
$$\underline{0 \cdot 2\,1\,3}$$
$$\underline{0 \cdot 5\,6\,9}$$

$0 \cdot 607 + 0 \cdot 8253$

$$0 \cdot 6\,0\,7\ \ +$$
$$\underline{0 \cdot 8\,2\,5\,3}$$
$$\underline{1 \cdot 4\,3\,2\,3}$$
$$1\ \ \ 1$$

Notice that the carrying (which may be remembered instead of written down) is done in exactly the same way as with whole numbers.

$0 \cdot 746 - 0 \cdot 314$

$$0 \cdot 7\,4\,6\ -$$
$$\underline{0 \cdot 3\,1\,4}$$
$$\underline{0 \cdot 4\,3\,2}$$

$0 \cdot 657 - 0 \cdot 148$

$$0 \cdot 6\ ^{4}\!\!\not5\ ^{1}7\ -$$
$$\underline{0 \cdot 1\ \ 4\ \ 8}$$
$$\underline{0 \cdot 5\ \ 0\ \ 9}$$

Notice the 'borrowing and paying back' is done in the same way as with whole numbers. This is because the figure to the left always has ten times the value.

One final example:

$0 \cdot 85 - 0 \cdot 637$

$$0 \cdot 8\,5\ \ \ \ -$$
$$\underline{0 \cdot 6\,3\,7}$$

This looks strange: we do not have anything to take the seven away from. The problem is solved by placing a zero after the five to make $0 \cdot 850$. Putting a zero in the hundredths column does not change the value of the number.

$$0 \cdot 8 \ ^4\!5 \ ^1 0 \ -$$
$$0 \cdot 6 \ \ 3 \ \ 7$$
$$0 \cdot 2 \ \ 1 \ \ 3$$

So far we have neglected to check our subtraction answers by addition. We must be confident that they are correct. The last one will be checked just to remind you how it is done.

$$0 \cdot 2 \ 1 \ 3 \ +$$
$$0 \cdot 6 \ 3 \ 7$$
$$0 \cdot 8 \ 5 \ 0 \quad \text{☑} \quad \text{☺}$$
$$1$$

MULTIPLYING

First we will try a simple multiplication of decimals.

$$0 \cdot 3 \ \times$$
$$0 \cdot 2$$

Clearly a 6 must be written in the answer space, but where should we place the decimal point?

To answer this we have a closer look at what the numbers mean. $0 \cdot 3$ means $3/10$ and $0 \cdot 2$ means $2/10$. We can now do this by multiplying the fractions, as shown.

$$\frac{3}{10} \quad \times \quad \frac{2}{10} \quad = \quad \frac{6}{100}$$

By NOT doing the normal cancelling we get the answer as a numbers of hundredths which tells us the 6 has to go in the hundredths column. Now if we follow the suggestion made at the start of this chapter and keep the decimal points in line the answer looks a little odd.

$$0 \cdot 3 \ \times$$
$$0 \cdot 2$$
$$0 \cdot 0 \ 6$$

We certainly do not want to go through this procedure every time so here is one way around the problem. If 0·3 is multiplied by 10 it becomes simply 3, and 0·2 × 10 = 2. Now we have multiplied the numbers in the sum by 10 × 10 which is 100. So the answer to 3 × 2 must be divided by 100 to give the answer to 0·3 × 0·2. This means that the 6 of our answer must be moved from the units space to the hundredths space which is 2 spaces to the right of the units. Remember the decimal point comes immediately to the right of the units to separate the units from the decimals. This can be done by moving the decimal place 2 spaces to the left. There is a lot in this paragraph and it is probably worth reading again.

So, 6 ÷ 10 = 0·6 and 6 ÷ 100 = 0·06

Our calculation is now set out in this way:

$$
\begin{array}{r}
0\cdot3 \ \times \\
\underline{0\cdot2} \\
\underline{0\cdot06}
\end{array}
$$

The calculation 3 × 2 is done and the answer 6 is written on the right without regard to the decimal point. Then the decimal point is inserted to make sure the 6 is in the hundredths space. Put another way, the number 0·3 has 1 place of decimals as does the number 0·2 and we have 1 + 1 = 2 places of decimals in the answer.

Suppose we wish to calculate 0·32 × 0·4. The first part of the multiplication is:

$$
\frac{4}{10} \quad \times \quad \frac{2}{100} \quad = \quad \frac{8}{1000}
$$

So multiplying tenths by hundredths will produce a figure in the thousandths place. This means that with the numbers shown we have two places of decimals in the first number and one place of decimals in the second number, and this results in three places of decimals in the answer.

It will always be the case that if we add the number of decimal places in the two numbers of the sum this is the number of decimal places in the answer.

Now we have an easier way of multiplying decimals. We do not need to keep track of the decimal point until we have the figures of the answer.

$$
\begin{array}{r}
0\cdot32 \ \times \\
\underline{0\cdot4} \\
\underline{0\cdot128}
\end{array}
$$

Had the above sum been 0.32×0.5 we would have had this result:

$$
\begin{array}{r}
0.32 \times \\
0.5 \\
\hline
0.160 \\
\hline
\end{array}
$$

Your calculator will give the answer to this sum as 0.16, which is perfectly correct and has only 2 places of decimals. Obviously this number is the same as 0.160. We must write this last zero in the answer to get the decimal point in the right place.

A more complicated example such as 0.384×0.59 would be set out as shown below:

$$
\begin{array}{r}
0.384 \times \\
0.59 \\
\hline
3456 \\
19200 \\
\hline
22656 \\
\hline
\end{array}
$$

To position the decimal point in the answer we must have $3 + 2 = 5$ decimal places, so the answer is 0.22656

DIVIDING

Working out where to place the decimal point in the answer when dividing by a decimal fraction is not easy, so we use a rather neat trick to simplify the problem.

We already know that the line in a fraction means divide, so 3/5 means $3 \div 5$. Also 30/50 means the same as 3/5.

Combining these two ideas we can say that $0.64 \div 0.4$ has the same answer as $6.4 \div 4$. All we have to do is divide 6.4 by 4.

$$
4\overline{)6.4}
$$

4 will go into 6 once with 2 left over. The 1 must be in the units space so the decimal point in the answer must go immediately after the 1.

$$
\begin{array}{r}
1. \\
4\overline{)6\,^24}
\end{array}
$$

Now 4 goes into 24 exactly 6 times so we can complete the answer.

$$\frac{1 \cdot 6}{4 \overline{)\, 6\ ^24}}$$

So the answer to $0 \cdot 64 \div 0 \cdot 4$ is $1 \cdot 6$.

It only remains to demonstrate what can be done when a decimal division does not work out exactly. The calculation of $0 \cdot 25 \div 0 \cdot 4$ is first changed to read $2 \cdot 5 \div 4$.

$$4 \overline{)\, 2 \cdot 5}$$

We first need to realise that $4 \times 6 = 24$, so $4 \times 0 \cdot 6 = 2 \cdot 4$. Refer back to the multiplication section if you are not sure about this. When we do this part of the division we have $0 \cdot 1$ left over and nowhere to put it. This is easily solved by putting a 0 in the next column.

$$\frac{0 \cdot 6}{4 \overline{)\, 2 \cdot 5\ ^10}}$$

Because of place value we can continue this sum without worrying about the decimal point any more. 4 goes into 10 twice with 2 left over so another zero is called for, and finally 4 goes into 20 exactly 5 times.

$$\frac{0 \cdot 6\ \ 2\ \ 5}{4 \overline{)\, 2 \cdot 5\ ^10\ ^20}}$$

Remember we can check that the answer to a division sum is correct by multiplying.

$$\begin{array}{r} 0 \cdot 6\, 2\, 5 \times \\ 4 \\ \hline 2 \cdot 5\, 0\, 0 \end{array} \quad ☑ \quad ☺$$

We now have a procedure for dividing by a decimal fraction. Both numbers have to be multiplied by 10 as many times as it takes to make the number we are dividing by into a whole number. Set out the sum with the decimal point above the decimal point in the number being divided. (This keeps all the figures in the correct places.) Then carry out the division as normal.

The final example follows, one step at a time.

$$0 \cdot 3604 \div 0 \cdot 53$$

Multiply both numbers by 100:

$$36 \cdot 04 \div 53$$

Set out the sum with the decimal point in the answer space:

```
        .
53)3604
```

53 will not go into 3 or 36 but it will go into 360. It takes two 50s to make 100, so six 50s will make 300. Now $53 \times 6 = 318$ and it looks as though 6 is the first part of the answer.

```
      0·6
53)3604
```

Take away 318 and bring down the next figure:

```
      0·6
53)3604
  318
  424
```

You will find that $53 \times 8 = 424$, so the answer can be completed.

```
      0·68
53)3604
  318
  424
  424
```

A check by multiplying 53 by 0·68 will show we have the correct answer.

CHAPTER FIVE

ARITHMETIC WITH PERCENTAGE FRACTIONS

My rent has gone up by twenty per cent.
Any higher and I'll live in a tent.

As in previous chapters, percentage fractions will be referred to as just percentages or per cent from here on. The sign for a percentage is %. So 38% is shorthand for 38 per cent. The word cent derives from hundred – 38% means 38 in every 100.

Just for completeness there is a related fraction called per mille, which means per thousand. The sign for this is ‰. This is so rarely used it will not be referred to again.

The actual sums involving percentages will, by now, be fairly straightforward. I am assuming that you are able to check any calculations shown. The difficulty is in making sure you are doing the correct calculation. What MUST be remembered is that a percentage is a number of hundredths of 'something'. I think the best approach is to show some examples of the use of percentages which you may come across.

Most calculations with percentages involve multiplying. If you are tempted to add percentages you can get some answers which are obviously wrong.

One example where adding is correct occurs in the run-up to elections where we are often told the results of opinion polls designed to predict the result. A result with fictitious names (I refuse to include politics in this book) could be shown as:

The 'Reduce Income Tax' party	24%
The 'Free Parking' party	41%
The 'Votes For Children' party	32%

These percentages are based on a sample designed to show the voting intentions of 100% of people eligible to vote. This is where adding is appropriate and these figures should add up to 100%. However, 24 + 41 + 32 = 97. The complete result should have been:

The 'Reduce Income Tax' party	24%
The 'Free Parking' party	41%

The 'Votes For Children' party	32%
Don't Know	3%

Rises in wages are often quoted in percentages, but most people are more concerned with how much extra money they will receive. If someone earning £380 per week was awarded a rise of 5% the first thing to realise is that this is 5 hundredths (5 hundredths of £380). It is easy to work out 1 hundredth of £380; it is £3·80 (move the decimal point 2 places to the left). So multiply this by 5 to get £19.

This calculation could have been put in the form of a fraction:

$$\frac{380 \quad \times \quad 5}{100}$$

This gives us a simple rule for finding a percentage of anything. We simply multiply the number by the percentage we want and then divide by 100. The answer is £19.

The new wage would then be £380 + £19 = £399. The more cautious would then think Income Tax has to be deducted. The current standard rate of Income Tax in the UK is 20%. The easy way to do this calculation is to realise that 10% is 1 tenth (10/100 = 1/10). 10% of £19 is £1·90. 20% is two times 10% so the Income Tax would be £3·80. The rise in wages is now reduced to £19 - £3·80 which is £15·20.

This leads on to working out your rise as a percentage of your old wage. It is best to investigate this in small steps. First we go back to simple questions and answers involving fractions.

What is 4 as a fraction of 8? It is clearly 1/2, but for the purpose of this explanation it is better to write this as 4/8.

What is this as a percentage? We want this fraction as a number of hundredths, so following the rules of fractions, multiply top and bottom by 100.

$$\frac{4}{8} \quad \times \quad \frac{100}{100}$$

If we do part of the working (4 × 100 ÷ 8 = 50) we get:

$$\frac{50}{100}$$

This is just what we want (a number of hundredths), so the answer is 50%.

This will surely work with any two numbers.

What is 9 as a fraction of 12? The answer is 9/12 (= 3/4). Multiply by 100/100.

$$\frac{9}{12} \quad \times \quad \frac{100}{100}$$

Work out $9 \times 100 \div 12 = 75$.

$$\frac{75}{100}$$

So 9 is 75% of 12.

Back to our percentage wage rise, after tax, we just change the figures in either of the previous examples to get $15 \cdot 20 \times 100 \div 380 = 4$. (It is quite permissible to cheat here and use a calculator – the main purpose is to understand what you are doing.)

This rule is usually stated slightly differently. To find one number as a percentage of another, divide the first number by the second and than multiply by 100. It does not matter which way you do it, you still get the same answer.

We now see that our 5% rise has been reduced to 4% after paying Income Tax. (Not so good if the inflation rate is 5%.)

Of course, to work out your real percentage increase, you would have to take into account National Insurance payments as well. I will leave this as an exercise for the reader.

You may have seen advertisements for car insurance with phrases like 60% discount if you have not had a claim in the last 5 years, a further 10% discount if there is only one driver and 15% off if you apply online.

This sounds marvellous. You appear to be getting a total discount of 85% (60 + 10 + 15). Yes it is too good to be true!

Let us suppose, taking into account type of car, driver's age, etc., the annual premium was £500. The first discount is calculated:

60% of £500 = £500 × 60 ÷ 100 = £300. The premium is reduced to £500 - £300 = £200.

The second discount is calculated:

10% of £200 = £20. (Not 10% of £500). The premium is reduced to £200 - £20 = £180.

The third discount is calculated:

15% of £180 = £27. The premium is reduced to £180 – £27 = £153.

If the percentages had been added first to give 85% then the premium would have been reduced by 85% of £500 = £425 and the final figure would have been £75.

Quite a difference. The above figures may still represent good value for money but you would have been slightly better off taking an insurance policy that offered a 70% discount for no claims in the last 5 years and no other discounts. (The discount would be 70 × £500 ÷ 100 = £350 resulting in a premium of £150.)

Sometimes when dealing with percentages the order is important. A good example is something we all moan about i.e. the price of petrol. At present there is a fuel duty of 58p added to the price of every litre of petrol and VAT at the current rate of 17·5% is also added to the basic price of about 43p per litre. There are two ways this calculation may be done. Checking that my VAT figures are correct would be good practice.

Fuel price	43p	Fuel price	43p
Fuel Duty	58p	VAT	7·5p (approx.)
VAT	17·7p (approx.)	Fuel Duty	58p
Total to pay	118·7	Total to pay	108·5

I will leave you to guess which method the government chooses!

Percentages are used in the calculation of VAT. The present rate of VAT is 17.5% - a rather awkward figure apparently. This may well be 20% by the time you read this but the figures below are still useful.
    If we have to add VAT to an article priced at £49·20 the calculation would be.

$$\frac{49 \cdot 2 \quad \times \quad 17 \cdot 5}{100}$$

This is where most people would reach for the calculator, but there is a little trick that makes it easy. (Mathematicians are always on the look-out for ways of avoiding hard work).
    The number 17·5 is equal to 10 + 5 + 2·5. So if we find 10% we can find half of that to get 5% and half again to get 2·5%. Add the three answers together and we have the required 17·5%. To find 10% (10/100 = 1/10) we move the decimal point one place to the left.

10% of £49·20 =	£4·92
5% of £49·20 =	£2·46 (half of £4·92)
2·5% of £49·20 =	£1·23 (half of £2·46)
17·5% of £49·20 =	£7·71 so total price is £49·20 + £7·71 = £56·91.

I have carefully chosen the price here so that the working is exact, but the method will work for any number. Just to prove this the VAT on £138·44 is worked out below:

10% of £138·44 =	£13·844
5% of £138·44 =	£ 6·922
2·5% of £138·44 =	£ 3·461
17·5% of £138·44 =	£24·227

Your calculator will confirm this is the correct answer. The nearest we can get to this is £24·23. Notice that if the thousandths column had been ignored we would have had the answer £24·22. (The Vatman would be most upset if this lower figure was used. More of this in the Chapter on accuracy. Total price is £138·44 + £24·23 = £162·67.

The next example using percentages is to find out the original price of an article before VAT was added. The unwary may think it is sufficient to subtract 17·5% from the price. Unfortunately, using the last example, 17·5% of £162·67 (use your calculator here) is £28·46725. Subtracting this is obviously not going to give us the right answer.

To find out what we should do it will be helpful to take a look at a type of question that appears in School text books under the heading of Ratio and Proportion. A typical question would be:

A car travelling at a constant speed travels 10 miles in 15 minutes. How long would it take to travel 28 miles?

This problem is related to fractions as both numbers can be multiplied or divided by the same number and still give a true statement. If both numbers were multiplied by 2 we would say the car travels 20 miles in 30 minutes. Again, dividing by 5 would mean the car travels 2 miles in 3 minutes. I hope this is obvious to you.

You may have already worked out the answer but the method shown here makes it easier to see what to do when the figures are not so easy.

10 miles corresponds to 15 minutes.

We have to rewrite this so that the 10 becomes 15. This is done in two stages. First change the 10 into a 1 by dividing by 10. Then change the 1 into a 28 by multiplying by 28.

Dividing by 10:

1 mile corresponds to 15/10 = 1·5 minutes.

Multiplying by 28:

28 miles corresponds to 28 × 1·5 = 42 minutes.

Notice the question required 10 miles to be changed to 28 miles so it is this that is reduced to 1 and then increased to 28.

If the question had asked how many miles would be travelled in 18 minutes then we have to change the 15 into 18. First dividing by 15, then multiplying by 18.

10 miles corresponds to 15 minutes.
$10/15 = 2/3$ miles corresponds to 1 minute.
$2/3 \times 18 =$ <u>12 miles</u> corresponds to 18 minutes.

I have gone into this example in great detail and kept the numbers simple because these types of problems very often occur in real life and can cause considerable confusion.

For the purposes of the VAT problem I will use a figure of 20%, partly because it keeps the numbers simpler and understanding the solution is the aim, not heavy Arithmetic, and also the VAT rate is likely to be 20% in 2011.

If something is priced at £85 before VAT this corresponds to 100% (100 hundredths).
The VAT to be added is 20% of £85 = £17.
Adding these together gives 120% of £85 = £85 + £17 = <u>£102</u>.
To get back from the final price to the price before VAT we can now set out the problem in the style used on the previous page remembering that we need to change from 120% to 100%:

120% corresponds to £102.
1% corresponds to £102/120 − £0·85.
100% corresponds to £0·85 × 100 = <u>£85</u>.

As we are multiplying the final price by the fraction 100/120 this can be simplified as multiplying by the fraction 5/6 will give the same answer.

The next example of careful use of percentages is posed as a question. If you were offered employment at a wage of £400 per week and you could choose whether to have an 8% rise every year or a 1·9% rise every quarter (13 weeks) which would you choose?

Most people may think that a rise of 8% per year is the same as a rise of 2% per quarter on the basis that $2 \times 4 = 8$. This reasoning leads to believing that a rise of only 1·9% per quarter is the worse option and so we would select a rise of 8% per year. WRONG CHOICE!

This is an example of comments made at the start of the chapter. A percentage is a number of hundredths of 'something'.

To save repetition the 1·9% per quarter rise will be called Option A and the 8% per year rise will be Option B.

At the end of the first quarter Option A wage goes up to £407·60 per week while Option B remains at £400. At the end of the second quarter Option A goes up to £415·34. 1·9% of £407·60 is bigger than 1·9% of £400. Poor old Option B stays at £400. At the end of the third quarter Option A earns £423·23. At the end of the fourth quarter Option A is now £431·27 and Option B at last gets a rise up to £432 per week. Option B has at last overtaken Option A but Option A has earned more money in the whole year.

The results over 2 years are summarised in the table: Don't worry about checking the calculations at this stage. This type of calculation will be explained in the Chapter on mixed numbers.

	Option A			Option B		
	Wages per qtr	Wages per year	Total for 2 years	Wages per qtr	Wages per year	Total for 2 years
Qtr 1	£5200·00			£5200·00		
Qtr 2	£5298·80			£5200·00		
Qtr 3	£5399·48			£5200·00		
Qtr 4	£5502·07	£21400·35		£5200·00	£20800·	
Qtr 5	£5606·61			£5516·00		
Qtr 6	£5713·14			£5516·00		
Qtr 7	£5821·69			£5516·00		
Qtr 8	£5932·30	£22973·74	£44374·09	£5516·00	£22464·00	£43264·00

After 2 years Option A has produced £1110·09 more than Option B and the difference increases every year! This problem alone should show a compelling reason for understanding Arithmetic.

The final example of the care needed when dealing with percentages is one that may have affected many readers.

If you were to take an examination in two parts and you gained a mark of 70% in the first part and 80% in the second part, you would naturally want to know your overall percentage mark. Adding is clearly not appropriate here as a mark of 150% is just as impossible as athletes declaring they will give 110% of their effort.

You make think that an average of 75% would be correct, but this is not necessarily so. More needs to be known about the NUMBER of marks awarded in each part.

If your actual mark in the first part was 35 out of 50 (equal to 70%) and in the second part was 160 out of 200 (equal to 80%) then your total mark would have been 195 out of 250, which corresponds to 78%

.

CHAPTER SIX

ARITHMETIC WITH NEGATIVE NUMBERS

Why do two minuses make up a plus?
Has it for ever and ever been thus?

In this chapter, to distinguish between positive and negative numbers in calculations they will be put in brackets with the appropriate sign. So (-5) indicates negative 5 and (+5) is positive 5 (just the ordinary number 5).

We start with a further look at the number line, introduced in Chapter 1.

Notice positive numbers indicate the space from 0 to the RIGHT and negative numbers indicate the space from 0 to the LEFT.

This immediately allows us to do the first calculation:

ADDING

$$(+4) + (-3)$$

We must start at +4 on the line and add the space corresponding to -3. This space goes to the left, so following the arrow on the line below we arrive at the answer.

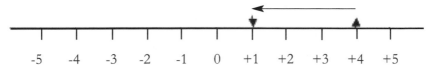

Our first sum is done (+4) + (-3) = (+1). Notice this gives the same answer as 4 − 3.

Another one:

$$(+3) + (-5)$$

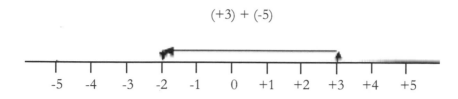

$(+3) + (-5) = (-2)$, exactly the same answer as $3 - 5$.

One more:

$$(-1) + (-3)$$

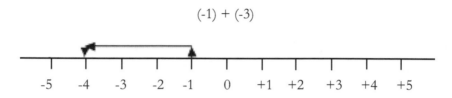

$(-1) + (-3) = (-4)$, exactly the same as $-1 - 4$.

This should be enough for you to see that no matter what the numbers are, adding a negative number will always produce the same result as subtracting.

If this is not enough to convince you, think of positive numbers representing the amount of money you have and negative numbers for the bills you have to pay.

If you have £87·90 in your bank account and a bill for £23·40 arrives in the post this negative number is ADDED to your bank account after payment leaving you with £87·90 - £23·40 = £64·50.

SUBTRACTING

We already know from Chapter 2 what happens when you subtract a positive number:

$$(+5) - (+3) = 5 - 3 = 2.$$

This allows us to say 'adding a negative number or subtracting a positive number is the same as subtraction'.

The only other type of sum to be dealt with here is subtracting a negative number. To understand this we will go back to the bill that arrived in the post. After receiving the bill (even before paying it) you only have £64·50 left to spend. Suppose now that the postman, on his way back, calls and says 'Sorry, I delivered that bill by mistake. I will take it back.' He has now SUBTRACTED that NEGATIVE number from you so your bank balance can now have £23·40 ADDED to it.

$$£64·50 - (-£23·40) = £64·50 + £23·40 = £87·90$$

If you are still not completely sure about this, remember in Chapter 2 subtraction sums were checked by adding. We will do it. The subtraction to be checked is:

$$
\begin{array}{r}
64·50 \ - \\
- \ 23·40 \\
\hline
87·90 \\
\hline
\end{array}
$$

If we add 87·90 and (-23·40) we should get 64·50.

$$
\begin{array}{r}
87·90 \ + \\
-23·40 \\
\hline
\\
\hline
\end{array}
$$

We know that adding a negative number is equivalent to subtracting, we now subtract:

$$
\begin{array}{r}
87·90 \ - \\
23·40 \\
\hline
64·50 \\
\hline
\end{array}
$$ ☑ ☺

We have now shown that subtracting a negative number is the same as adding.

MULTIPLYING and DIVIDING

As explained in an earlier chapter this process is simply repeated addition so we should expect similar results.

$$(+5) \times (-8)$$

Here we are adding the number (-8) five times and we will get the answer (-40).

Multiplication gives the same answer when we reverse the numbers.

So (-8) x (+5) must also give the answer (-40).

This will work with all numbers and this tells us that if we multiply a positive number and a negative number the answer is negative.

To understand multiplying two negative numbers we need to go back a little to repeated addition. When we add three fours together it is very simple. However addition has to have a starting point, which is zero (this is so obvious with positive numbers it is not considered worth mentioning). So we could consider the sum set out like this:

$$
\begin{array}{r}
0 + \\
4 + \\
4 + \\
\underline{4 +} \\
12
\end{array}
$$

If we try (-3) x (-4) the question is how do we add (-4) to 0 minus three times? Very tricky. To make it easier we will look at (+1) x (-4). This means adding 0 and (-4) to get (-4). If we now move on to (-1) x (-4) this must be reverse of adding 0 and (-4), in other words subtracting (-4) from 0 and we now know the answer to this is (+4). If we repeat this three times we get the answer (+12)

$$
\begin{array}{r}
0 - \\
-4 - \\
-4 - \\
\underline{-4 -} \\
+ 12
\end{array}
$$

We now have the rather surprising result; multiplying two negative numbers gives a positive answer. Surprising, but the only thing that fits the pattern and makes sense.

To understand division with negative numbers we will look at the sum:

$$(+8) \div (-2)$$

We need to know how many (-2) s will make (+8). Obviously we can keep on adding (-2) s together and we will never arrive at (+8). But there must be an answer. Bearing in mind what happened with multiplication we can guess that the answer may be (-4). (In spite of what you may have been told there is nothing wrong with guessing the answer and then testing to see if it is correct.)

$$(+8) \div (-2) = (-4) \,?$$

Now to check the answer, if (-4) x (-2) = (+8) then we are right. Sure enough, it works, so:

$$(+8) \div (-2) = (-4) \quad ☑ \quad ☺$$

If we try:

$$(-8) \div (+2) = (-4) \,?$$

and test in the same way this is also correct.

Finally:

$$(-8) \div (-2)$$

It seems logical that the answer will be (+4). The test is to multiply (+4) by (-2) and we do get the answer (-8), so:

$$(-8) \div (-2) = (+4) \quad ☑ \quad ☺$$

These sums show that if we divide a positive number and a negative number (either way round) the answer is a negative number, but if both numbers are negative then the answer is positive.

The results of this chapter can be summarised in this table:

	+	-
+	+	-
-	-	+

This is often remembered as 'a plus and a minus make a minus and two minuses make a plus'. This is a useful memory aid as long as you understand the results.

CHAPTER SEVEN

ARITHMETIC WITH MIXED NUMBERS

Whole numbers and fractions mixed up together.
Will this make me feel under the weather?

A mixed number is a combination of a whole number, which may be positive or negative, with any type of fraction.

Examples are $3^1/4$, $8\cdot375$ and 158%. The percentage example is a mixed number because 100% represents 100 hundreds, which is the same thing as 1.

We will first deal with whole numbers and fractions.

ADDING

$$3\tfrac{1}{4} \ + \ 2\tfrac{1}{2}$$

This means the same as:

$$3 \ + \ \tfrac{1}{4} \ + \ 2 \ + \ \tfrac{1}{2}$$

We can do the adding in any order we choose, and is easier to add the whole numbers first. At the same time the bottom numbers can easily be made the same: refer back to Chapter 3 if you are not sure. The sum can now be written as:

$$5\ \frac{1 \quad + \quad 2}{4}$$

And the final answer is:

$$5 \ \frac{3}{4}$$

If we change the above sum slightly we can deal with the only other problem likely to occur when adding these numbers:

$$3 \ \frac{3}{4} \ + \ 2 \ \frac{1}{2}$$

This becomes:

$$5 \ \frac{3 \quad + \quad 2}{4}$$

and then:

$$5 \ \frac{5}{4}$$

5/4 is an improper fraction (see Chapter 3) and should be written as 1¹/4 and then added to the 5. So the final answer is:

$$6 \ \frac{1}{4}$$

SUBTRACTING

Once again we will start with a simple example:

$$3 \ \frac{3}{4} - 2 \ \frac{1}{2}$$

As before the whole numbers can be subtracted and the bottom numbers made the same:

$$1\frac{3\quad-\quad2}{4}$$

The answer is clearly:

$$1\frac{1}{4}$$

As with adding there is only one problem likely to arise and this involves some careful borrowing and paying back.

$$7\frac{1}{5}\quad-\quad2\frac{3}{10}$$

The first stage, as above is:

$$5\frac{2\quad-\quad3}{10}$$

We now have the problem of working out 2 – 3. The answer, of course, is -1, but this is no help here. We need to borrow and the only place we can borrow from is the 5 units. What we do is to borrow 1 leaving 4 in the units and this 1 has to be paid back in the form of 5 fifths. (This is equivalent to borrowing a £5 note from a friend and paying your friend back with 5 £1 coins.)

$$4\frac{7\quad-\quad3}{10}\quad=\quad4\frac{4}{10}=4\frac{2}{5}$$

Sometimes it is not so obvious that this is going to happen as in the next sum:

$$2\frac{3}{5}-1\frac{2}{3}$$

First stage:

$$1 \quad \frac{9 \quad - \quad 10}{15}$$

Here we must borrow from the units leaving nothing. The fraction part is a number of fifteenths so we must pay back 15 of these, so the sum becomes:

$$\frac{24 \quad - \quad 10}{15} \quad = \quad \frac{14}{15}$$

MULTIPLYING

Start with a simple sum again:

$$1 \, \frac{3}{4} \times 2 \, \frac{1}{2}$$

If we attempt what seems obvious we would multiply 1 by 2 getting 2 and multiply 3/4 by 1/2 making 3/8 and give the final answer as $2^3/8$. This demonstrates a very important part of Arithmetic and that is to see when we have got a wrong answer.

If we multiply 3/4 by 2 we get 6/4 which can be written as $1^1/2$, so $1^3/4$ multiplied by 2 must be $2 + 1^1/2 = 3^1/2$. The correct answer to our sum cannot be smaller than this.

The correct way to do this calculation is to change both fractions to improper fractions and then follow the rule established in Chapter 3.

There are 4 quarters in 1 unit and we must add a further 3 to write the first fraction as 7/4. In the same way 2 units is the same as 4 (2 × 2) halves and we have 1 more to add making 5/2 for the second fraction.

The sum is now written as:

$$\frac{7}{4} \quad \times \quad \frac{5}{2}$$

There is no cancelling possible so we multiply the top numbers and multiply the bottom numbers as before.

$$\frac{7}{4} \quad \times \quad \frac{5}{2} \quad = \quad \frac{35}{8}$$

This is an improper fraction and should be changed back to a mixed number in the following way: $35 \div 8 = 4$ ($4 \times 8 = 32$) with a remainder of 3, so the answer is best written as:

$$4 \; \frac{3}{8}$$

One more example to make sure that you can get every stage right.

$$3 \; \frac{1}{5} \times 2 \; \frac{3}{4}$$

Change both numbers to improper fractions ($3 \times 5 + 1 = 16$) and ($2 \times 4 + 3) = 11$.

$$\frac{16}{5} \quad \times \quad \frac{11}{4}$$

Do the cancelling. Both 16 and 4 can be divided by 4 giving 4 and 1.

$$\frac{4}{5} \quad \times \quad \frac{11}{1}$$

Multiply the top numbers ($4 \times 11 = 44$) and multiply the bottom numbers ($5 \times 1 = 5$).

$$\frac{44}{5}$$

Change back to a mixed number ($44 \div 5 = 8$ with a remainder of 4). So the answer is:

$$8 \; \frac{4}{5}$$

DIVIDING

This process has to be done in a similar way to multiplying. First the mixed numbers are changed to improper fractions. Then we use the previously established rule of turning the second fraction upside down and then multiplying. Finally, if we are left with an improper fraction we change it back to a mixed number.

$$6 \frac{2}{5} \div 1 \frac{1}{15}$$

Change to improper fractions ($6 \times 5 + 2 = 32$) and ($1 \times 15 + 1 = 16$).

$$\frac{32}{5} \div \frac{16}{15}$$

Turn the second fraction upside down and change to multiplying...

$$\frac{32}{5} \times \frac{15}{16}$$

Doing the cancelling ($32 \div 16 = 2$, and $15 \div 5 = 3$) we can complete the sum.

$$\frac{2}{1} \times \frac{3}{1} = \frac{6}{1} = 6$$

All the work for dealing with mixed numbers using decimals has already been done so just four examples should be enough as a reminder.

When adding and subtracting the decimal points must be kept in line.

$$7 \cdot 5 \, 2 \, 9 + 1 \cdot 3 \, 6 \, 5 \, 3$$

$$\begin{array}{r} 7 \cdot 5 \, 2 \, 9 \quad + \\ 1 \cdot 3 \, 6 \, 5 \, 3 \\ \hline 8 \cdot 8 \, 9 \, 4 \, 3 \end{array}$$

$$5 \cdot 6 \, 5 \, 3 - 3 \cdot 1 \, 4 \, 6$$

$$
\begin{array}{r}
5 \cdot 6 \; {}^{4}\!5\,{}^{1}3 - \\
3 \cdot 1 \;\; 4 \;\; 6 \\
\hline
2 \cdot 5 \;\; 0 \;\; 7
\end{array}
$$

When multiplying the decimal point is not needed in the working.

$$2 \cdot 6 \, 1 \, 5 \times 1 \cdot 5 \, 8$$

$$
\begin{array}{r}
2 \cdot 6 \, 1 \, 5 \times \\
1 \cdot 5 \, 8 \\
\hline
2 \, 0 \, 9 \, 2 \, 0 \\
1 \, 3 \, 0 \, 7 \, 5 \, 0 \\
2 \, 6 \, 1 \, 5 \, 0 \, 0 \\
\hline
4 \cdot 1 \, 3 \, 1 \, 7 \, 0
\end{array}
$$

There are a total of 5 figures after the decimal points in the question so we must have 5 places of decimals in the answer which is 4·1 3 1 7. (The 0 at the end has no effect.)

$$8 \cdot 7 \, 4 \div 2 \cdot 3$$

Multiply both numbers by 10 to get 8 7 ·4 ÷ 2 3.

$$
\begin{array}{r}
3 \cdot 8 \\
2 \, 3 \,) \, \overline{8 \, 7 \cdot 4} \\
6 \, 9 \\
\hline
1 \, 8 \, 4 \\
1 \, 8 \, 4 \\
\hline
\end{array}
$$

Any percentage greater than 100% is a mixed number. However, percentage fractions can be mixed with other fractions, for example 68·25% or $33^{1}/3\%$.

Probably the most important type of calculation with these numbers is to find the size of a number when it is increased by a percentage. Referring back to Chapter 5, the calculation of the new wage when £380 is increased by 5% could be done in one sum.

£380 corresponds to 100% of the wage and to get the new wage we have to work out 105% of £380.

$$\frac{380 \quad \times \quad 105}{100}$$

It is easy to divide by 100 and it does not matter whether we divide 380 or 105 by 100. It turns out to be more useful to divide 105 by 100 to get 1·05.

So, the only sum to do is £380 × 1·05 which is £399, as found previously. This will work for all percentage increases. A few examples are given in the following table.

Percentage increase	Multiply by
1	1·01
9	1·09
10	1·10
25	1·25
50	1·50
90	1·90
99	1·99
100	2·00

There remains just one more type if calculation to be considered. An example would be: a quantity has increased from 15 to 60, what is this as a percentage?

The increase here is 45. As a 'fraction' of the original amount this is 45/15 and to convert this to a percentage it must be multiplied by 100.

$$\frac{45}{15} \times 100$$

This gives an answer of 300%. Perhaps surprisingly, this shows that when the original quantity is multiplied by 4 the increase is 300% (not 400% as you might expect).

CHAPTER EIGHT

RECURRING DECIMALS

Do these numbers go on for ever,
And ever and ever and ever?
My word that's clever.

Many decimal division sums do not work out exactly. One of these was shown in Chapter Two: 79 ÷ 3. We had to leave the answer as 26 with a remainder of 1. One way of completing the sum is to say that the remainder of 1 still needs to be divided by 3 and this, of course, can be written as 1/3 and so the answer can be given as:

$$26 \ \frac{1}{3}$$

There is another way of continuing this sum by writing 79 as 79·00:

$$\frac{2 \ 6 \cdot 3 \ 3}{3 \,)\, 7 \ ^1 9 \cdot ^1 0 \ ^1 0}$$

Clearly this pattern is not going to change and so the answer could be written as:

$$26 \cdot 33\ldots..$$

The dots after the last 3 indicates that the number 3 will carry on for ever.

This is called a recurring decimal and is more usually written as $26 \cdot \dot{3}$ and here the dot over the three indicates that the 3 carries on for ever.

Another example of dividing by 3 gives a rather different result:

$$\frac{2 \ 6 \cdot 6 \ 6}{3 \,)\, 8 \ ^2 0 \ ^2 0 \ ^2 0}$$

The reason for these results when dividing by 3 is that if an exact answer is not obtained then the only possible remainders are 1 or 2. So if we have to add zeros to continue the division we either arrive at $10 \div 3$ or $20 \div 3$. The first of these leads to a series of threes and the second to a series of sixes.

Dividing by many other numbers produces similar results. For example $20 \div 11$.

$$\begin{array}{r} 1 \cdot 8\ 1\ 8\ 1 \\ 1\ 1\ )\ 2\ 0\ ^9 0\ ^2 0\ ^9 0\ ^2 0 \end{array}$$

This answer will continue repeating the numbers 1 and 8.

This answer is often written as $1 \cdot \overset{\cdot}{8} \overset{\cdot}{1}$, the dots appearing over the first and last numbers of the repeating sequence.

Dividing by 7 produces a longer series of repeating numbers using all possible remainders as in $15 \div 7$.

$$\begin{array}{r} 2 \cdot 8\ 5\ 7\ 1\ 4\ 2\ 8\ 5\ 7 \\ 7\ )\ 1\ 5\ ^6 0\ ^4 0\ ^5 0\ ^1 0\ ^3 0\ ^2 0\ ^6 0\ ^4 0\ ^5 0 \end{array}$$

The sequence of numbers 8, 5, 7, 1, 4 and 2 can only go on repeating and the answer is usually written as:

$$2 \cdot \overset{\cdot}{8} 5\ 7\ 1\ 4 \overset{\cdot}{2}$$

The recurring dots again over the first and last numbers in the repeating sequence. Notice the recurring dots are only used on numbers after the decimal point.

These recurring decimals are more common than you might think. The reason is that only numbers made up of multiples of only 2 or only 5 or both will produce exact answers.

Our number system is based on 10 and these strange answers are all because of the fact that 3, 6, 7 and 9 cannot give exact answers when divided into 10.

CHAPTER NINE

EQUALITY AMONG FRACTIONS

Two-quarters and six-twelfths were having a fight
Over who was biggest and best.
Three-fifths came along and got into the mix
Saying you're both point five and I am point six.

Sometimes it is useful to be able to change from one type of fraction to another and this can now be done quite easily.

Changing a vulgar fraction to a decimal fraction is done by dividing the top number by the bottom number. As was seen in the last chapter this may result in a recurring decimal. Just two examples should be sufficient. The first is 12/25.

```
         0·4 8
25)1 2 0 0
   1 0 0
     2 0 0
     2 0 0
```

The second is 17/24.

```
         0·7 0 8 3
24)1 7 0 0 0 0
   1 6 8
     2 0 0
     1 9 2
       8 0
       7 2
        8
```

This is where the 3 in the answer will carry on repeating so this is written as:

$$0 \cdot 7 \ 0 \ 8 \ 3$$

The problem of comparing fractions with different denominators mentioned on Page 3 can now be solved. The fractions should be converted to decimals first.

Vulgar fractions are changed to percentage fractions by multiplying by 100 and again two examples should make this clear.

$$\frac{3}{4} \quad \times \quad \frac{100}{1} \quad = \quad \frac{300}{4} \quad = \quad 75\%$$

$$\frac{5}{7} \quad \times \quad \frac{100}{1} \quad = \quad \frac{500}{7} \quad = \quad 71 \ \frac{3}{7} \ \%$$

Changing a decimal fraction to a vulgar fraction is slightly more difficult but fortunately there is an easy short cut to be found. We will start with 0·75 as we should already know the answer. The 7 is in the tenths column and the 5 is in the hundredths column so we can write this as:

$$\frac{7}{10} \quad + \quad \frac{5}{100}$$

Making the bottom numbers the same:

$$\frac{70 \quad + \quad 5}{100} \quad = \quad \frac{75}{100}$$

Both the top and bottom numbers can be divided by 25 so the answer is:

$$\frac{3}{4}$$

If you look at the pattern of the numbers in the fraction before it was reduced the top number is the number after the decimal point and the bottom number is 1 followed by a 0 for each number of the decimal fraction. You should see that this will work for any decimal. So:

$$0 \cdot 37 \quad = \quad \frac{37}{100}$$

and:

$$0 \cdot 048 \quad = \quad \frac{48}{1000} \quad = \quad \frac{6}{125}$$

To change a decimal fraction to a percentage fraction we also multiply by 100, which means moving the decimal point 2 places to the right. For example, $0 \cdot 375 = 37 \cdot 5\%$.

A percentage can be changed to a decimal in the opposite manner (moving the decimal point 2 places to the left). $94\% = 0 \cdot 94$ and $3\% = 0 \cdot 03$.

A percentage fraction is changed to a vulgar fraction by dividing by 100 and using the simplest version of the fraction. Examples are:

$$50\% \quad = \quad \frac{50}{100} \quad = \quad \frac{1}{2}$$

and:

$$7 \cdot 5\% \quad = \quad \frac{7 \cdot 5}{100} \quad = \quad \frac{3}{40}$$

The results of this chapter are summarised in the table showing the most commonly occurring equivalents.

EQUALITY AMONG FRACTIONS

Vulgar Fraction	Decimal Fraction	Percentage Fraction
$\dfrac{1}{10}$	0·1	10%
$\dfrac{1}{8}$	0·125	12·5%
$\dfrac{1}{5}$	0·2	20%
$\dfrac{1}{4}$	0·25	25%
$\dfrac{1}{3}$	0·$\dot{3}$	33¹/3%
$\dfrac{3}{8}$	0·375	37·5%
$\dfrac{2}{5}$	0·4	40%
$\dfrac{1}{2}$	0·5	50%
$\dfrac{3}{5}$	0·6	60%
$\dfrac{5}{8}$	0·625	62·5%
$\dfrac{2}{3}$	0·$\dot{6}$	66²/3%
$\dfrac{7}{10}$	0·7	70%
$\dfrac{3}{4}$	0·75	75%
$\dfrac{4}{5}$	0·8	80%
$\dfrac{7}{8}$	0·875	87·5%
$\dfrac{9}{10}$	0·9	90%

81

CHAPTER TEN

BRACKETS AND ORDER OF OPERATIONS

Numbers in brackets first set the trend
Then comes times and divide
Plus and minus are set to one side
To await their turn at the end.

The word operation refers to any way of combining numbers (addition, subtraction, multiplication and division).

If we have a sum with two or more operations we can get different answers depending on which way we do it. We have already come across one such sum in Chapter 2.

$$2 \times 3 - 3 \times 2$$

We can do this is in the order it is written, $2 \times 3 = 6$, $6 - 3 = 3$ and then $3 \times 2 = 6$.

Another way is to reverse the order and get the same result.

We could do the subtraction first, $3 - 3 = 0$, $2 \times 0 \times 2 = 0$.

We could do multiplying first and then subtraction, $2 \times 3 = 6$, $3 \times 2 = 6$, $6 - 6 = 0$.

The way to overcome this problem is to have a universal agreement about which order is correct. The order chosen is that all multiplication and division must be done before attempting addition and subtraction. If it is necessary to change this order for any reason then the calculations to be done first are enclosed in brackets.

A simple example of this would be if you were paying the bill for 4 people at a restaurant where any starter cost £3, any main course cost £11 and a sweet was £4. To work out the total cost if the 4 people had all three courses the sum is 3+11+4×4. If we followed the above rule we would get the wrong answer and we must insist that, in this case, the adding is done first. This is shown by using brackets and working out anything in brackets first – (3+11+4) × 4 – giving the answer 72. In reality this sum would be written as:

$$4(3+11+4)$$

If the contents of a bracket are to be multiplied by a number it is normal to put the number in front of the bracket and omit the multiplication sign (the reason for leaving out the multiplication sign will be explained in a later Chapter). Unfortunately, if your calculator can use brackets you will have to put the multiplication sign back. This sum would have to be entered on a calculator as $4 \times (3 + 11 + 4) =$.

Our rule now has to be modified to read 'do everything in brackets first, then the multiplying and dividing and finally adding and subtracting'. A short word has been devised to make this easy to remember and this is BODMAS. The O stands for of, which was used earlier in connection with multiplying, and is inserted to make the word pronounceable. The meanings of the other letters should be obvious.

The order in which multiplying and dividing is done does not affect the answer, in some countries the word used is BOMDAS.

One word of warning about brackets is that they are not always obvious. When the top and bottom parts of a fraction have yet to be worked out and a long line is used (we have seen this in adding and subtraction) this line acts as a bracket. So:

$$\frac{9 \quad + \quad 7}{8}$$

must be worked out as $(9 + 7) \div 8 = 16 \div 8 = 2$.

A further point to note about the restaurant bill example is that we will also get the right answer if we work out $4 \times 3 + 4 \times 11 + 4 \times 4 = 12 + 44 + 16 = 72$. It is important to understand that multiplying or dividing something in brackets by a number affects every addition and subtraction inside the brackets. It is reasonable to translate this as 'multiplying and dividing are more powerful than adding and subtracting'. So:

$8(7 - 3 + 12)$ means the same as $8 \times 7 - 8 \times 3 + 8 \times 12 = 56 - 24 + 96 = 128$

and:

$$\frac{24 \quad - \quad 36 \quad + \quad 72}{6}$$

means the same as $24 \div 6 - 36 \div 6 + 72 \div 6 = 4 - 6 + 12 = 10$.

We should now be able to work out fairly complicated pieces of Arithmetic. In the example below I have kept the numbers simple. With more awkward numbers a calculator would be used but you still have to know what you are doing.

$$\frac{35 \ (\ 18 \ - \ 4 \ \times \ 3 \ + \ 5 \ )}{2 \ + \ 5}$$

Brackets first and multiplying inside bracket:

$$\frac{35 \ (\ 18 \ - \ 12 \ + \ 5 \ )}{7}$$

$$\frac{35 \ \times \ 11}{7}$$

Now we can cancel and divide top and bottom by 7 giving:

$$5 \times 11 = 55.$$

CHAPTER ELEVEN

PRIME NUMBERS AND FACTORS

You may try many times
To find factors of primes
But they do not exist
So you'll have to desist.

Contained in the set of whole numbers there is another important set of numbers which can only be divided by itself or one. For example 37. Only the numbers 1 and 37 will divide exactly into 37. These numbers are called prime numbers. With only this definition we would have to call 1 a prime number, but this is regarded as trivial, and so is not included.

The first few prime numbers are 2, 3, 5, 7, 11, 13, 17 and 19. Please feel free to reach for your calculator and find some more.

Mathematicians have spent years trying to find a formula which will predict prime numbers but there is no pattern that can be followed.

At this stage I must introduce the mathematical meaning of the word 'product'. It is just the technical name for the answer to a multiplication sum and its use saves a lot of writing.

It is a rather surprising fact that every number can be written as the product of prime numbers in one, and only one way. This turns out to be very useful. Just one of the ways it can help is when adding and subtracting fractions where we wish to find the smallest common multiple of two or more numbers.

The way to find these numbers is to repeatedly divide by the smallest possible prime number at each stage. To do this it is easier to have each answer underneath the sum so that it is available for immediate use without setting out another division sum. A simple example should make this clear. We will start with the number 60. The smallest prime number that will divide exactly into 60 is 2.

$$2 \overline{)6\,0}$$
$$3\,0$$

This can be repeated:

$$2 \overline{)6\,0}$$
$$2 \overline{)3\,0}$$
$$\qquad 1\,5$$

We cannot divide by 2 again but we can divide by 3 (the next prime number):

$$2 \overline{)6\,0}$$
$$2 \overline{)3\,0}$$
$$3 \overline{)1\,5}$$
$$\qquad 5$$

5 is a prime number so we cannot go any further. If you remember how multiplication was used to check the answer to a division sum you will see we can now say:

$$60 = 2 \times 2 \times 3 \times 5$$

These numbers 2, 2, 3 and 5 are called the prime factors of 60.

The division has been shown in stages, but in practice it would be done as one continuous calculation:

$$2 \overline{)6\,0}$$
$$2 \overline{)3\,0}$$
$$3 \overline{)1\,5}$$
$$5 \overline{)\;\;5}$$
$$\qquad 1$$

I have included here the final stage to give the last answer as 1. Some people prefer to do this last stage and then write down the product shown by the numbers in the first column.

The Arithmetic is usually very simple even when dealing with large numbers, say, 74970.

$$2 \overline{)7\,4\,9\,7\,0}$$
$$3 \overline{)3\,7\,4\,8\,5}$$
$$3 \overline{)1\,2\,4\,9\,5}$$
$$5 \overline{)\;\;4\,1\,6\,5}$$
$$7 \overline{)\;\;\;\;8\,3\,3}$$
$$7 \overline{)\;\;\;\;1\,1\,9}$$
$$17 \overline{)\;\;\;\;\;1\,7}$$
$$\qquad\qquad 1$$

$$74970 = 2 \times 3 \times 3 \times 5 \times 7 \times 7 \times 17.$$

Expressing numbers as a product of prime numbers allows a fairly easy way to find smallest common multiples for adding or subtracting fractions. Suppose we had three fractions to deal with having bottom numbers of 18, 24 and 60. If we were to multiply these numbers together we would get 25920 (not nice). Instead we will work out the prime factors of each number.

$$
\begin{array}{r}
2\overline{)1\,8} \\
3\overline{)9} \\
3\overline{)3} \\
1
\end{array}
$$

$$18 = 2 \times 3 \times 3$$

$$
\begin{array}{r}
2\overline{)2\,4} \\
2\overline{)1\,2} \\
2\overline{)6} \\
3\overline{)3} \\
1
\end{array}
$$

$$24 = 2 \times 2 \times 2 \times 3$$

And we already know:

$$60 = 2 \times 2 \times 3 \times 5$$

A common multiple of 18, 24 and 60 must have the following:
    three factors of 2 (otherwise it cannot be a multiple of 24):
    two factors of 3 (otherwise it cannot be a multiple of 18):
    and a factor of 5 (otherwise it cannot be a multiple of 60).
So the smallest common multiple of 18, 24 and 60 is $2 \times 2 \times 2 \times 3 \times 3 \times 5 = 360$.
    If we had to apply this to a fraction sum the numbers are now much less daunting.

$$\frac{7}{18} \quad + \quad \frac{5}{24} \quad - \quad \frac{11}{60}$$

We want all the bottom numbers to be 360. It might not be easy to see how many 18s there are in 360. In the prime factors of 360, if we cross out the ones already used for 18, the remaining factors must produce the answer.

$$2 \times 2 \times 2 \times 3 \times 3 \times 5$$

We are left with $2 \times 2 \times 5 = 20$. So the top number 7 must be multiplied by 20.
If we do the same for 24:

$$2 \times 2 \times 2 \times 3 \times 3 \times 5$$

We are left with $3 \times 5 = 15$ and the 5 must be multiplied by 15. I think it is fairly easy to see that $60 \times 6 = 360$. The fraction sum is now:

$$\frac{140}{\ } + \frac{75}{360} - \frac{66}{\ } = \frac{149}{360}$$

There is a further use for these prime factors. There are times when it is necessary to know what is the largest number that will divide exactly into all of several numbers. As we already have the prime factors of 18, 24 and 60 we will find out the largest number that will divide exactly into all of them.

$$18 = 2 \times 3 \times 3$$
$$24 = 2 \times 2 \times 2 \times 3$$
$$60 = 2 \times 2 \times 3 \times 5$$

2×3 is the only combination of factors that appears in each set of factors, so 6 is the largest number that will divide exactly into 18, 24 and 60.

   This is a good place to introduce a piece of mathematical shorthand. If a number is multiplied by itself several times the writing can be shortened by using an index number placed after the number and in a higher position. Using this method 5×5 is written as 5^2 showing that two fives are to be multiplied. If the index is 2 this is said as '5 squared, or 5 to the power of two'. If the index was three we would say '5 cubed, or 5 to the power of three'. For indices (the plural of index) greater than 3 we would say 5 to the power of the index. In this shortened form we can write:

$$24 = 2^3 \times 3 \qquad \text{and} \qquad 360 = 2^3 \times 3^2 \times 5$$

CHAPTER TWELVE

ACCURACY AND APPROXIMATION

Approximation
The sun is about a hundred million miles away,
That's good enough for me.

Accuracy
I'll pay you about ten pounds an hour said he,
You won't say I. The rate is twelve forty three.

Most of the operations described in this book are guaranteed to have exact answers. The exception to this is division (apart from dividing fractions). We have seen with recurring decimals how division can result in answers that are never ending and we are then forced into a decision of when to stop writing. If we are dealing with money in pounds the answer is apparently simple: stop after the second decimal place (which represents pence) as we do not have a coin smaller than 1p.

If we were dividing £14·46 by 9 we would get £1·60666... and this would have to be written either as £1·60 or £1·61. To decide which, we need to be aware of the fact that 1·605 is exactly half way between 1·60 and 1·61, so 1·60666 is clearly closer to 1·61 and our answer would have to be £1·61. Even though this answer is not exact, it is described as accurate to 2 decimal places.

If, in a money calculation, we had arrived at an answer of £1·605, we have a problem as £1·605 is equally far away from £1·60 as it is from £1·61. It has to be generally agreed that in this case we round up to £1·61. As long as everyone does this we will always agree on the answer.

It seems that if we need to 'lose' one place of decimals then if it is 1, 2, 3 or 4 we simply ignore it, but if it is 5, 6, 7, 8 or 9 then the previous figure must be increased by 1.

Naturally it is not quite this simple! Suppose our answer is 1·4449 and we decide 3 places of decimals is enough then we write 1·445. We then change our minds and go for 2 decimal places and write 1·45. There is something wrong here as 1·4449 is nearer to 1·44 than it is to 1·45 (just). This means our rule of the last paragraph must be amended to read 'if we are losing a 4, look one more place ahead to be sure'.

Just one more example of rounding to 2 decimal places. If we had an answer of 3·497 we cannot increase the 9 so the answer has to be written as 3·50. The zero here is important as it shows that the number is correct to 2 decimal places. 3·5 would only guarantee accuracy to 1 decimal place.

I mentioned in the last paragraph that 1·4449 is nearer to 1·44 than it is to 1·45 (just). The curious reader may ask 'how close can you get to 1·45'? The best we could do would appear to be 1·4449... where the 9s carry on for ever. This is rather a difficult concept. It means that if you spent the rest of your life writing 9s, then passed the task on to your children and grandchildren there would still not be enough nines. (Warning. Do not try this at home.) However there is a rather smart trick we can use to answer the question.

If we multiply 1·4449... by 10 we get 14·449.... If we then do the following subtraction:

$$10 \times 1{\cdot}4449\ldots = 14{\cdot}44999999\ldots\ -$$
$$\underline{\ 1 \times 1{\cdot}4449\ldots = \ \ 1{\cdot}44499999\ldots}$$
$$9 \times 1{\cdot}4449\ldots = 13{\cdot}005000\ldots$$

As the nines go on for ever there will only be a series of 0s after 13·005. If we now divide 13·005 by 9 we get 1·445. We now have the rather surprising conclusion that 1·4449... is exactly equal to 1·445. (We have also done some algebra!)

When doing large numbers of similar calculations a marvellous tool is a computer spreadsheet. Very briefly a spreadsheet is usually in the form of a table. Each cell (this is the technical name for a box) in the table can contain numbers, letters or instructions. These instructions are often about Arithmetic; for example, add up the numbers in other specified cells. Spreadsheets are often used for calculations involving money, so 2 places of decimals is sufficient, but care has to be taken to get this rounding done in the right way.

Suppose someone is starting a business involving buying component parts, assembling them and then selling the finished product. For simplicity we will take the case of just six products. These will be known as products A, B, C, D, E and F.

In the business plan our entrepreneur has decided that the cost price of the materials for each item must be increased by 75% to make his (or her) business profitable.

The following table shows some of the records for a week. I hope it is self-explanatory.

Product	Cost Price(£)	Selling Price(£)	Number Sold	Money Received(£)
A	37.59	65.78	7	460.48
B	58.19	101.83	13	1323.82
C	28.19	49.33	9	443.99
D	44.59	78.03	11	858.36
E	85.89	132.81	9	1195.27
F	12.69	22.21	11	244.28
	Total Received(£)			4526.20

The instruction in the Selling Price column is 'multiply Cost Price by 1·75 (this is the easy way of increasing a number by 75%). The instruction in the Money Received column is 'multiply Selling Price by Number Sold' and the instruction in the Total Received row is 'add up all the Money Received'.

Those of you who like to check answers will see that there is something wrong here. A calculator will give the answer 460·46 when 65·78 is multiplied by 7. A further check by calculator will show the rest of the answers in the money received to be 1323·79, 443·97, 858·33 and 244·31. Has my computer made a mistake? The answer to this is an almost certain NO! Actual computer mistakes are so rare that we can forget that possibility. The mistake is in the instructions given to the computer (this is almost always the case when the excuse 'computer error' is given).

What has happened here is that the computer was instructed to DISPLAY the shaded money figures to 2 decimal places. This is what a lot of people do when designing money spreadsheets. However, the computer is too clever for us here. The correct answer to 37·59 × 1·75 is 65·7825. The computer displays this as 65·78, as instructed, but remembers the correct answer for the next calculation. So the Money Received becomes 65·7825 × 7 = 460·4775, displayed as 460·48.

In this case the instructions in the highlighted cells must insist that the figures are rounded to 2 decimal places before the next calculation is done. This results in the correct spreadsheet, shown next.

Product	Cost Price(£)	Selling Price(£)	Number Sold	Money Received(£)
A	37.59	65.78	7	460.46
B	58.19	101.83	13	1323.79
C	28.19	49.33	9	443.97
D	44.59	78.03	11	858.33
E	85.89	132.81	9	1195.29
F	12.69	22.21	11	244.31
	Total Received(£)			4526.15

You may think that an error of 5 pence in the total is not much, but if hundreds of different items were made and sales figures were in the thousands this would be a substantial error.

Whenever numbers arise from a measurement, exactness is not possible and we have to decide what would be a sensible level of accuracy. When measuring lengths with a metre rule most people would be happy to measure to the nearest millimetre (abbreviated to mm). So if a length was measured as 374mm we would be confident that it must be between 373·5mm and 374·5mm otherwise it would have been rounded to 373mm or 375mm. We could then say this length is 374mm plus or minus 0·5mm. There is a mathematical sign for this and we can write 374mm ± 0·5mm.

Some care is needed when calculations are done with measurements. If the area of a rectangular garden was needed, perhaps to find out how much grass seed was required to make a lawn, we may decide the area would be accurate enough to the nearest square metre. With the length and width measured as 20 metres and 15 metres the area would be calculated as 20 × 15 = 300 square metres. However the length and width could have been as small as 19·5 metres and 14·5 metres, leading to an area of 19·5 × 14·5 = 282·75 square metres, or as large as 20·5 metres and 15·5 metres, resulting in an area of 20·5 × 15·5 = 317·75 square metres. Clearly not good enough! Also, if the perimeter was needed to find the length of edging required this could lie between 68 metres and 72 metres.

This is not the place for repeated calculations so I will leave the reader to see what the results would be if the lengths were measured to the nearest 0·1 metre. The two measurements would then be 20·1 ± 0·05 and 15·1 ± 0·05 metres. The answers may surprise you.

We now come to approximation, arguably one of the most important skills needed in Arithmetic. This is where the original numbers in a calculation are replaced by simpler numbers in order to get a rough idea of the answer.

This should always be done before attempting a particularly difficult calculation which would normally be done on a calculator. An example would be: 38·4 × 79·85. This could be approximated to 40 × 80 = 320. If my calculator shows a supposedly correct answer of 303749·4 I know something is very wrong. (In fact I entered 3804 by mistake instead of 38·4). The sensible approximation prevents acceptance of the wrong answer. The more complicated the numbers the more important it is to approximate first.

Approximation is often important when converting from one set of units to another. For example, 1 mile is exactly the same as 1·609344 km. Rather a difficult calculation to do if you are driving when you want to change 145 miles into kilometres. Now a reasonable approximation would be 1 mile ≂ 1·6 km. (The sign ≂ means approximately equal to). But we can do better, 1·6 = 16/10 = 8/5. We can now change from miles to kilometres if we multiply by 8 and divide by 5 (or the other way round if it is more convenient). 145 ÷ 5 = 29 and 29 ×8 = 232. The exact answer is 233·35488. I think 232 km is good enough for most people. Naturally, to change from kilometres to miles the number of kilometres is multiplied by 5 and that answer is divided by 8.

Another popular conversion between units is that between litres and gallons. There are still motorists around who like to know how many gallons of petrol or diesel fuel they have purchased and the pumps only show litres. The accurate conversion is 1 gallon is the same as 4·54596 litres. If we use 4·5 as an approximation the conversion becomes fairly easy as 4·5 can be written as 9/2. So to change from gallons to litres we would multiply by 9 and divide by 2. The more usual conversion would be to change litres to gallons by multiplying by 2 and dividing by 9. Picking a nice easy figure of 27 litres this would become 6 gallons. A more accurate answer is 5·94 gallons, probably close enough for most purposes

There are too many more examples of useful approximations to be listed here and of course the accuracy required would depend on individual requirements.

CHAPTER THIRTEEN

ALTERNATIVE NUMBER SYSTEMS

Binary

Just the two of us, noughts and ones,
and we are expected to do everything.

The number system in everyday use today is based on the number 10, presumably because we have 10 digits (fingers and thumbs) on our hands. This has not always been the case. (The number base, not the fingers and thumbs). For example the Babylonians had a system based on 60. This makes for very difficult Arithmetic. The remnants of this system are still around and it is why we have sixty seconds to the minute and sixty seconds to the hour.

To do any Arithmetic involving times of day requires the use of the 24 hour clock. For any of you not completely familiar with this a time of 9.30am would be written as 09:30 and 7.45pm would be written as 19:45. All pm times have 12 added to the hours. The abbreviations am and pm are from Latin words meaning before midday and after midday, so in spite of what you may read occasionally there is no such time as 12am or 12pm. At midday 12 o'clock is called 12 noon or 12:00 and midnight is called 12 midnight or 24:00. If you need to do any formal Arithmetic involving times of day remember two things. The first is when adding; if the minutes column adds up to more than 60 then 1 is carried to the hours column. Also when subtracting, if 1 is borrowed from the hours then 60 must be paid back to the minutes.

There is no reason why we should be restricted to 10 as the base for our number system. To show how any other number system would work we need to examine carefully what we do at present. In the base 10 number system we have just 10 symbols (0, 1, 2, 3, 4, 5, 6, 7, 8 and 9). When we have counted up to 9 the units column is full. The units column is then restarted at 0 and a 1 is added to the column to the left making 10. This rule is constantly repeated as we carry on counting. So the counting can continue moving through 10, 11,... 19, 20,...29, 30,...99, 100, 101, etc. Using the idea of an index introduced in Chapter 11 we can write our column headings of thousands, hundreds, tens and units as 10^3, 10^2, 10 and units. This pattern of indices continues as far as we wish so the ten thousands column is 10^4.

The above paragraph may seem rather detailed. Its importance lies in the fact that if we change the base of our numbers the above rules must still apply.

The number system to be explained uses the number 2 as the base (also known as the binary system). It is important as it is how a computer does its work and it incidentally leads us to a very easy way of multiplying. The binary system uses only the symbols 0 and 1. Using the rules, counting is shown in the table below, together with the 'normal' equivalent numbers.

Binary	0	1	10	11	100	101	110	111	1000	1001	1010
Decimal	0	1	2	3	4	5	6	7	8	9	10

Following the previous pattern the column headings in binary would be 2^4, 2^3, 2^2, 2 and units.

Binary Arithmetic is as easy as it could get. The complete addition and multiplication tables are shown below.

+	0	1
0	0	1
1	1	10

×	0	1
0	0	0
1	0	1

The purpose of this book is not to teach binary Arithmetic, but if any readers have developed an interest in Arithmetic it can be quite instructive to do simple calculations in one system, then change the numbers to the other system and check that your answers are the same.

An example of the promised easy way of multiplying follows. The numbers to be multiplied are placed at the head of two columns. On the next row the first number is divided by 2 and any remainder is ignored and the second number is multiplied by 2. This process continues until the first number is reduced to 1. Then, in the second column cross out all the numbers opposite an even number and add up the rest to get the answer.

Divide by 2	Multiply by 2
107	84
53	168
26	~~336~~
13	672
6	~~1344~~
3	2688
1	5376
Add up	8988

Just for amusement, we will change 107 and 84 into binary and try the multiplication using the simple binary Arithmetic to see what has happened.

To change a number from base 10 (properly known as denary) to binary we first divide by 2 and note the remainder. This remainder tells you how many units there are (this can only be 0 or 1). Divide by 2 again and the remainder is the number of 2s. Continue doing this as long as you can.

$$
\begin{array}{r r l}
2) & 1\,0\,7 & \\
2) & 5\,3 & r\,1 \\
2) & 2\,6 & r\,1 \\
2) & 1\,3 & r\,0 \\
2) & 6 & r\,1 \\
2) & 3 & r\,0 \\
2) & 1 & r\,1 \\
& 0 & r\,1 \\
\end{array}
$$

The binary number is shown in the remainders, reading from bottom to top. 107 in the decimal system is therefore 1101011 in binary.

This is usually shortened to read $107_{10} = 1101011_2$

In the same way it can be shown that $84_{10} = 1010100_2$

The multiplication is set out in the usual way but only multiplication by 0 and 1 is required.

$$
\begin{array}{r}
1\,0\,1\,0\,1\,0\,0 \times \\
\underline{1\,1\,0\,1\,0\,1\,1} \\
1\,0\,1\,0\,1\,0\,0 \\
1\,0\,1\,0\,1\,0\,0\,0 \\
0\,0\,0\,0\,0\,0\,0\,0\,0 \\
1\,0\,1\,0\,1\,0\,0\,0\,0\,0 \\
0\,0\,0\,0\,0\,0\,0\,0\,0\,0\,0 \\
1\,0\,1\,0\,1\,0\,0\,0\,0\,0\,0 \\
\underline{1\,0\,1\,0\,1\,0\,0\,0\,0\,0\,0\,0} \\
\underline{1\,0\,0\,0\,1\,1\,0\,0\,0\,1\,1\,1\,0\,0} \\
\end{array}
$$

To change this number back to a denary number we must work out $2^{13} + 2^9 + 2^8 + 2^4 + 2^3 + 2^2 = 8192 + 512 + 256 + 16 + 8 + 4 = 8988$. The Arithmetic here could not be easier, but we pay the price of the numbers rapidly becoming too long.

The connection to the 'easy multiplying method' becomes clearer in the repeat of the binary sum below. The numbers on the right show the denary equivalent of each row of the calculation.

```
        1 0 1 0 1 0 0 ×
        1 1 0 1 0 1 1
        1 0 1 0 1 0 0              84
       1 0 1 0 1 0 0 0            168
      0 0 0 0 0 0 0 0 0             0
     1 0 1 0 1 0 0 0 0 0          672
    0 0 0 0 0 0 0 0 0 0 0           0
   1 0 1 0 1 0 0 0 0 0 0          2688
  1 0 1 0 1 0 0 0 0 0 0 0         5376
 1 0 0 0 1 1 0 0 0 1 1 1 0 0
```

CHAPTER FOURTEEN

INTRODUCTION TO ALGEBRA

I am unknown and of neither sex
My closest friends just call me 'x'

This chapter has been included in an attempt to show that Algebra is not the fearsome subject that many think it is. In Arithmetic we are given a question involving some known numbers and we are required to find the answer. In elementary Algebra we are often given the answer and part of the question and the task is then to work out the rest of the question.

About the simplest problem in Algebra is: 'I am thinking of a number. I then add 5 to the number and get the answer 12. What was the number I thought of?'

I am sure you have all worked out the answer to be 7.

To ask the above question used a lot of words and Mathematicians like everything to be written in the shortest way possible. This is where the frightening letter 'x' starts to be used and most schoolchildren lose their interest. The use of the letter x is purely historic and if you wish use another letter or even a short word there is nothing wrong with this provided you make it clear what you are doing. Using x to stand for 'the number thought of' the short way of stating the above problem is:

$$x + 5 = 12$$

Emphasis on the meaning of the '=' sign is needed here. It means that whatever is to the left of the sign is EXACTLY the same as whatever is to the right, and when we use the sign as shown we form what is called an equation. If we do the same thing to both sides of an equation we will get another equation where both sides are EXACTLY the same. It is quite permissible to add 3 to both sides of the above equation making the further equation:

$$x + 5 + 3 = 12 + 3$$

shortened to:

$$x + 8 = 15$$

The last two equations are perfectly correct but they do not help us to find the answer. What we need to do is to subtract 5 from each side, as follows:

$$x + 5 - 5 = 12 - 5$$

becomes

$$x + 0 = 7$$

and finally

$$x = 7$$

and we have solved our first problem using Algebra.

The steps in solving our equation were explained in great detail in order to make it clear that we had to 'undo' the process of adding 5 by subtracting 5. This idea of 'undoing' is nearly always used when solving simple equations.

If we had the equation:

$$x - 8 = 9$$

we then have to 'undo' the process of subtracting 8 by adding 8.

$$x - 8 + 8 = 9 + 8$$

leading to:

$$x - 0 = 17$$

and the answer:

$$x = 17$$

Once this idea is understood the solution can be considerably shortened.

For example:

$$x + 14 = 7$$

Subtract 14 from both sides:

$$x = -7$$

I hope you remember negative numbers from Chapter 6.

There remains one more possibility with this type of equation, shown here:

$$7 - x = 3$$

It seems that all we can do here is to add x to both sides and see if it helps.

$$7 - x + x = 3 + x$$

Whatever x is equal to, $- x + x$ can only be zero, so:

$$7 = 3 + x$$

Subtracting 3 from both sides leaves us with $4 = x$, which the same as $x = 4$.

We can now look at equations involving multiplying and dividing. There is another piece of shorthand first. If we write $3 \times x$, meaning 3 multiplied by x, it could easily be mis read as $3xx$, so the multiplication sign is omitted and we write just $3x$ to mean 3 times x. (This idea was first mentioned in Chapter 10). As the multiplication sign is the ONLY sign that is ever left out this should not cause any confusion. Now to solve our next equation:

$$3x = 12$$

As x has been multiplied by 3 we must undo this by dividing both sides by 3.

$$\frac{3x}{3} = \frac{12}{3}$$

When we have divided both 3x and 12 by 3 we have our answer:

$$x = 4$$

This technique will clearly work with any numbers, for example:

$$4 \cdot 5x = 9$$

If we divide both $4 \cdot 5x$ and 9 by $4 \cdot 5$ we arrive at the answer:

$$x = 2$$

A further variation would be:

$$\frac{8}{x} \quad = \quad 6$$

We must now multiply both sides by x to get:

$$8 = 6x$$

A final division of both sides by 6 tells us the answer is $x = 8/6$.

This answer can be simplified and written as:

$$\frac{4}{3}$$

or:

$$1\tfrac{1}{3}$$

The final stage in this Chapter is to combine the previous ideas to solve slightly more difficult equations.

$$2x + 1 = 7$$

The easiest first stage here is to subtract 1 from both sides resulting in:

$$2x = 6$$

Then divide both sides by 2 giving the answer:

$$x = 3$$

There is an alternative way of solving equations of this type. This is to draw what is called a flow diagram indicating what has been done to x, shown below:

To 'undo' this we must start on the right of the diagram and 'undo' each stage in turn to get back to the beginning.

Further work with Algebra will not be done here. I hope you can now see that Algebra is not quite as bad as you thought.

To complete this introduction to Algebra we will solve a practical problem relating to the lawn measurement in Chapter 12. Suppose we have bought 70 metres of lawn edging and we want the length to be 3 metres more than the width. How wide is the lawn? Just for a change we will use w to stand for the width of the lawn, so the length will be $w + 3$.

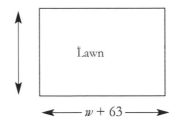

The equation we can write is:

$$w + w + 3 + w + w + 3 = 70$$

We have four ws and two 3s on the left so we can write:

$$4w + 6 = 70$$

Subtracting 6 from both sides:

$$4w = 64$$

Divide both sides by 4 and the answer is:

$$w = 16$$

So, the width of the lawn is 16 metres and the length is 19 metres, and if we add two lengths and two widths we have the correct answer of 70 metres.

PUZZLES

1. This puzzle is to make all the numbers from 1 to 10 using exactly four 4s. You are only allowed to use the signs for adding, subtracting, multiplying, dividing and brackets. You may also use 44 if necessary. For example $(4 \times 4)/(4 \times 4) = 1$.

2. Think of any 3 digit number (492 for example); repeat it as a 6 digit number (492492). Enter the 6 digit number into your calculator, divide it by 13, divide the answer by 11 and divide that answer by 7. Can you explain the answer? A hint will be found in Chapter 2 – Multiplying.

3. How many squares are there on the playing surface of a chess board (8 squares by 8 squares)? The answer is much more than 64 as the small squares can be grouped together to form larger squares.

4. If a clock takes 5 seconds to strike 6 o'clock, how long does it take to strike 12 o'clock? Of course it is not 10 seconds!

5. What is the exact value of 99·9… (99·9 recurring)? A hint will be found on page 98.

6. A similar puzzle is what is the exact value of $1/2 + 1/4 + 1/8 + 1/16 + 1/32 + \ldots\ldots$?

7. Rather a difficult one until you spot the trick. Can you find an easy way to add up all the whole numbers from 1 to 1000?

8. This one is not arithmetic but it should make you think. In the middle of a circular pond 100 metres across is a vertical pole 10 metres tall. A rope needs to be attached to the pole without throwing the rope or entering the pond. You can use up to 500 metres of rope. How is it done?

ANSWERS

Puzzle 1

$1 = (4 + 4)/(4 + 4)$

$2 = 4/4 + 4/4$

$3 = (4 + 4 + 4)/4$

$4 = 4 + (4 - 4)/4$

$5 = (4 \times 4 + 4)/4$

$6 = (4 + 4)/4 + 4$

$7 = 44/4 - 4$

$8 = 4 + 4 + 4 - 4$

$9 = 4 + 4 + 4/4$

$10 = (44 - 4)/4$

There are other possible answers to some of these.

Puzzle 2

The answer lies in the fact that $13 \times 11 \times 7 = 1001$, so if a 3 digit number is multiplied by 1001 the digits are repeated. If you set out the sum it will become obvious.

Puzzle 3

There is 1 big square surrounding the whole board.
A 7 by 7 square can be seen in 4 positions (top left, top right, bottom left and bottom right, making another 4 squares.
6 by 6 squares can be seen in 9 positions, 5 by 5 squares in 16 positions, 4 by 4 squares in 25 positions, 3 by 3 squares in 36 positions, 2 by 2 squares in 49 positions and finally the 64 small squares.

So the answer is 1 + 4 + 9 + 16 + 25 + 36 + 49 + 64 = 204.

Puzzle 4

It is the gaps between the hour strikes that take up the time and when a clock strikes 6 o'clock there are 5 gaps between strikes so each gap lasts for 1 second.

When the clock strikes 12 o'clock there are 11 gaps between the strikes so the time taken is 11 seconds.

Puzzle 5

$10 \times 99 \cdot 9999999..... = 999 \cdot 9999999.....-$
$\underline{\ 1 \times 99 \cdot 9999999..... = \ 99 \cdot 9999999.....,}$
$\ \ 9 \times 99 \cdot 9999999.... = 900 \cdot 000.......$

So $99 \cdot 999..... = 100$ exactly.

Puzzle 6

$2 \times \text{Answer} = 1 + 1/2 + 1/4 + 1/8 + 1/16 + ...-$
$\underline{1 \times \text{Answer} = \ \ \ \ \ 1/2 + 1/4 + 1/8 + 1/16 + ...}$

So the answer is 1.

Puzzle 7

Answer =   1 +   2 +   3 +   + 998 + 999 + 1000
Answer = 1000 + 999 + 998 +   + 3 +   2 +   1

Adding the two versions of the answer gives:

$2 \times$ Answer = 1001 + 1001 + 1001 +   +1001 + 1001 + 1001.

There are 1000 numbers altogether, each equal to 1001, so the total must be 1001 × 1000 = 1001000. This equals twice the answer we want, so the total of all the numbers from 1 to 1000 = 1001000 ÷ 2 = 500500.

Puzzle 8

Tie a loop in one end of the rope. Lay the rope around the outside of the pond. Put the free end through the loop and pull.